Henry L. Roth

**A Sketch of the Agriculture and Peasantry of Eastern Russia**

Henry L. Roth

**A Sketch of the Agriculture and Peasantry of Eastern Russia**

ISBN/EAN: 9783337298708

Printed in Europe, USA, Canada, Australia, Japan

Cover: Foto ©berggeist007 / pixelio.de

More available books at **www.hansebooks.com**

# A SKETCH

OF THE

# AGRICULTURE AND PEASANTRY

OF

# EASTERN RUSSIA.

BY

## HENRY LING ROTH.

LONDON:

BAILLIÈRE, TINDALL, & COX, KING WILLIAM ST., STRAND;

[ PARIS, AND MADRID. ]

1878.

# PREFATORY REMARKS.

THIS little book is compiled from Notes taken during a two-years' stay in the Province of Samara, in Eastern European Russia. The greater part of the time was spent in practical farming at Timashevo, a village situated about sixty miles east of Samara town ; but before my departure I travelled through the neighbouring country for a few months, visiting whatever districts I had heard to be interesting. I made also some acquaintance with the agriculture of Kazan, Orenburg, Simbirsk, and Saratoff; but of the two latter Provinces, and of Cis-Volgian agriculture in general, I know so little that I have refrained from mentioning anything about it. A glance at a map will show that the whole extent of country over which this Sketch ranges is very large, approaching in size the area of Ireland.

Notwithstanding the many difficulties I had as a stranger, with an imperfect acquaintance with the Russian language, to encounter in collecting my notes, I have always endeavoured to obtain the most correct information, and to do so have always been in the habit of repeating my questions over and over again, and to various individuals, before I ventured to note down the answers ; for even without any intentional prevarication there is always contradictory evidence on all matters. However, should the reader find any inaccurate statement I shall be happy to stand corrected.

The many gentlemen I visited, and to whom I feel deeply indebted for their generous hospitality and kindliness, are far too numerous for me to thank individually. I must, therefore, restrict myself to this short expression of my gratitude. They one and all received me with the greatest courtesy, and were most willing to help me to obtain all the information I required.

What with the hospitality of the Russians, and the great work of improvement that lies open to the energetic agriculturist, I felt more sorry at parting from Russia than with any other country I have yet visited.

The town of Samara has become illustrious through the introduction of Koomiss-Cure by Dr. Postnikoff.

I have also visited the German and Swiss Colonies on the banks of the Volga, but the information I obtained about them is too meagre to be recorded.

In the orthography of all Russian words and proper names, I have endeavoured to adhere as closely as possible to their pronunciation in the vernacular.

H. L. ROTH.

48, WIMPOLE STREET, LONDON.

*July, 1878.*

# CONTENTS.

# AGRICULTURE AND PEASANTRY

OF

# EASTERN RUSSIA.

―-ᴇᴧᴇ-

## CHAPTER I.

### THE FEATURES OF THE COUNTRY.

THE country is not interesting from an æsthetical point of view; it has no scenery. It is either all forest or all plain, and presents none of that happy variety and blending of trees, houses, hedges, and gardens that makes English landscape so pleasant. It lacks greatly in liveliness.

The internal communication is very bad. I have driven several hundred miles in a springless cart across the Steppes, as the dry, open, treeless fields and prairies are called, and have always found travelling unpleasant and uncomfortable. Roads, so to speak, there are none ; there is certainly a beaten track from village to village, but it is a mere strip of land, originally, I suppose, marked out as the shortest between two points, but now as irregular uneven, and as uncared-for as possible.

B

There are no ditches to carry off the water, which stagnates sadly to the hindrance of the traveller ; but in Russia one is not particular. If the track is too bad, then one drives through the adjoining fields, crop or no crop. Others follow in the new track formed ; until at last, in consequence of constant deviations to right and left, as circumstances require, the road becomes all askew, and at places nearly a quarter of a mile in breadth.

The bridges are entrusted to the peasants, who contract for repairs, which are seldom carried out. A traveller prefers at all times to ford a river than to cross a bridge ; for not only is the former method the safer course to pursue, but generally the mud lies so thick fore and aft the bridge that the great question arises how to get at it without sticking fast.

In travelling, one meets with little to enliven the journey, unless it be the everlasting strings of creaking waggons, laden with grain, and embellished by a sleeping peasant, who lies at full length on the top of each. Occasionally the traveller will meet gangs of men, or bevies of women, journeying in search of work. Greetings are then exchanged, caps raised on both sides ; and when they have passed wondering on, the traveller is again left alone with the jingling, tinkling bells, and the incomprehensible conversation carried on by the driver to the horses. Presently a tarantass (open, springless carriage) whirls rapidly past, with dashing nags, followed by huge clouds of black, blinding dust. Later on, a scowling wolf, startled by the horses' clatter, will swiftly cross the path, or an eagle, disturbed in his solitude, rises heavily on the traveller's approach. Rooks and crows, cawing delightedly over the carcase of some newly-fallen bullock, or magpies enjoying the last delicate morsels from the well-cleaned bones of some poor old horse, flights of bustards and the whistling sooslick, are all passed in turn.

In the meanwhile, the sun is blazing overhead with tropical severity, the heat is intense, the traveller's throat is dry and parched, yet there are no signs of water. The waving grain on the one hand, and the dry pasture on the other, are all he sees, let him strain his eyes to the utmost; and so, for miles and miles, as if it would go on for ever in eternal monotony and dust. At last, down a hollow in some dried up river-bed, he spies a pretty little lake, which welcomes him with the cool shade of its thriving willows. As he draws nigh, the shrill cry of the lapwing in its spasmodic flight greets him ; a splashing is heard from the water, and a brace of ducks or a flock of cranes rise high, to disappear in distant dark-blue space.

After a while he approaches the village, whose church spire has long loomed in view. Hobbled horses neigh at grass, stray frisky calves are passed, and then the village is entered where the nags are

to be changed. Luckily it is dry weather, otherwise the series of dung and rotten straw heaps by
which it is surrounded would let the cart sink in
axle-deep. Swine, poultry, and children fly to right
and left as the traveller drives up, followed by the
furious barking and growling of savage dogs, who
appear to have a special dislike to the bells dangling
above the horses' heads. If the women are at home,
sashes will fly open, and inquisitive faces may be dis-
cerned peering behind the dingy windows. The
houses (log-huts, cabins) of the peasantry all stand in
a row in the villages, forming two or three long and
broad streets, often half a mile in length. The
neighbouring land - owner's (and ancient serf - proprietor's) house, outwardly little better looking than
those of the peasantry, is situated at one end of the
street; and the white-washed village church, with its
green, pear-shaped cupolas, at the other end,—the
middle of the street being occupied by a series of
diminutive granaries. The tall posts, with their long
levers for drawing water from the wells, stand con-
spicuously to the front.

At the Posting-house the Samovar (national tea-
urn) is soon set to boil; several glasses of scalding
tea are gulped down; axles are greased; and before
the traveller is well seated, and has lighted his
cigarette, the fresh, restless horses are galloping down
the street. As he leaves the village, the kine already
stand lowing for admission in front of their respective

A RUSSIAN VILLAGE IN THE PROVINCE OF SAMARA.

yards, while the good housewives call and chuckle to the timid bleating lambs. Night is falling fast. It is getting very cold; the driver puts on his sheepskin, and the traveller doffs his overcoat for a nice warm fur, in which he can nestle like a bird in nest. The same dull monotony has to be gone through again. Few human beings are met with; and as light grows dimmer, nature grows calmer, the wild birds' last cry has softened into silence, the last sooslick has retired to rest, — the jingling bells and the horses' tramp are the sole disturbers of the peace. The traveller lies comfortably back in his tarantass, deep in a pleasant reverie. Suddenly, with a strong jerk, the horses pull up, the driver utters a cry, and the cart comes to a stand-still! The driver had fallen asleep, and has just awakened to find himself on the wrong track. He should have arrived at his destination long ago, and now he knows not where he is. This admission of the unfortunate man is met with violent abusive censure and sarcasm, notwithstanding his servile and dejected excuses. In the meanwhile clouds have gathered overhead, it is pitch dark. To search for the lost track is useless, they know well enough the search would last till day-break. So the horses are unharnessed, and their fore-legs shackled; the driver rolls himself up under the cart, and is soon fast asleep; while the traveller, after smoking a few cigarettes to the hooting and whirring of the owls above, and the munch, munch of the grazing nags, soon follows the example set by his unfortunate' servant.

The land in Russia is not enclosed. Ring-fences, indeed fences of any description, are, if not unknown, at least not to be seen in practice. The boundaries of estates or of peasants' lands are defined by terminal posts; the imaginary straight line drawn between two consecutive posts marking the limits of the adjoining lands. Frequently rivulets or ravines form a natural boundary, and occasionally a ditch and bank are the substitutes for fences. Ditch and bank are, however, very inadequate for the confinement of live stock. When a public thoroughfare runs through an estate, the ditch and bank become totally useless; for in wet weather it is so frequently impossible to wade through the swampy state of the road in low position, that the traveller, prince or peasant, drives through the ditch and over the mound (which is easily done with the small carts), and then continues his journey through the adjoining fields. Moreover, the effects of the atmosphere on the banks made of such friable material as black soil are so great that they soon crumble down, and then the temptation to over-ride them is the greater. The absence of fences, however, is the more inexplicable because of the disadvantages both landlords and peasants thereby labour under. The temptation to the poor or dishonest peasants to let their cattle trespass is great; quarrels frequently arise as to whether the cattle crossed the boundary or not, and if not amicably settled, they often lead to retaliative robbery and incendiarism.

The estate of Mr. Boolighin, at Krotoffka, forms an only and solitary exception in being enclosed : the whole of this gentleman's estate was enclosed with a rough paling at a considerable expense. Great honour is due to him as having set the example to his fellow proprietors.

The dissolving of the snow in spring, and the accompanying ice-floes on the rivers, are phenomena special to Russia. When the ice and snow commence to thaw, bridges are taken down, and sluices opened to ease the passage of the waters. The quantity of water that flows down is enormous. It increases daily. Rivulets are formed on the hills, which rush down with terrific force, and tear the soft soil with them, doing a vast amount of damage to the cultivated land they may happen to traverse. In low-lying parts, where perhaps during the winter much snow was driven by the storms, lakes are now formed, marshes are replenished ; and there where the traveller passed safely in summer, he will be astonished to find an impassable river, impassable on account of its rapid current and great depth. As the snow continues to dissolve, the water increases ; lakes and marshes over-flow into the nearest rivers, and these again, terribly swollen, overstep their ordinary limits to such an extent that the land is nearly everywhere under water. The streets in whole villages are so many canals; low-lying houses are swamped; and from others the water is only kept out by surrounding them plentifully with dung and earth from the yards. The mills on the river banks are as specks on the ocean. The water will rise ten to twenty feet above its natural level; and while it is at its height I need hardly add that the inhabitants of villages are completely cut off from all communication except by means of boats.

When the waters have disappeared many weirs are seen to have disappeared also; but the damage done lies principally in the manner in which the rivulets cut into the soil. The water, while running along or down a declivity, appears to dig up suddenly any soft spot not well covered with turf; occasionally, if there is grass growing thickly immediately beyond, then no evil results; but otherwise the whole spot is scooped out, and continues annually to enlarge, until after a few years often a large and deep ravine has been formed. I know of such a ravine at Markoff, where within the last nine years the water has worked back some thirty feet, and to a depth of twenty. As a ravine deepens it enlarges also, the sides fall in, and cause a retardation of the running water; then less earth is transposed, plants begin to grow, and soon it has all the appearances of an old hollow.

Owing to the absence of trees on the Steppes some

of the rivers cease to exist by the middle of summer, —their broad beds being occupied either by a tortuous thread of trickling water, an occasional stagnant, stinking pool, or are otherwise perfectly dry. Water can, therefore, only be retained by dams thrown across the beds, and planted with willows. The artificial little lakes thus formed seldom dry up altogether.

These lakes are often the only reservoirs for miles and miles, and without them neither man nor beast could exist. Ravines on hill-sides are treated in the same way; they are generally stocked with tench and a species of carp, which thrive amazingly.

In the Stavropol, Boogoolma Boogoorooslan, and Samara districts, where forest is still plentiful, the rivers do not dry up, at least not to such an extent as lower south, and are in consequence used for driving mills. It is remarkable, that in these mills each pair of stones is driven by a separate water-wheel. The mills are for the greater part the property of the nobles (landed proprietors), who let them to the peasants. A good mill, with three to four pair of stones, may be worth 1500 to 2000 roubles (£190 to £250) per annum. The miller's fees vary from one-ninth to one-twelfth of the quantity of corn ground. The dams thrown across the river are huge rude compositions of faggots, branches, and saplings, interlarded with straw, dung, turf, and loose clay earth ; their stability is not very great, and fre-

quent repairs are necessary, owing to the rotting away
of the faggots, etc., which being placed parallel with
the current, allow the water to trickle through the
whole breadth of the dam unseen. A breach is thus
soon made, and the following spring the greater part
is washed away.

---

## CHAPTER II.

### THE CLIMATE.

THE climate, partaking naturally of a continental
character, is an extreme one. The lengthened winter
causes farming operations to be restricted to the six
months of the year, beginning about the third week
in April, and ending approximatively with the first
week in October. The snows have barely begun to
dissolve (early in April) before the trees and shrubs
commence to bud, and the weather is for a time warm
and genial. The inundations, however, caused by the
dissolving snow and breaking-up of the ice, render it
still impossible to get on to the greater part of the
fields ; and at the same time the temperature decreases considerably, in consequence of the great
bodies of water about. As soon as these waters sub-
side, operations begin, and go on until October, when
early frosts and snows put an end to autumnal
ploughing.

The climate is droughty in so far as regards the agricultural half of the year, though on the whole the complaint is rather the scarcity of rain at the right time—*i. e.*, in the spring—than in a general absence of rain, the rainfall in autumn being copious enough. As instances which happen not seldom, I may mention that in some parts of the province of Samara in 1876, there was no or insufficient rain in spring until too late to do any good. The same was the case last year in Kazan, where I saw oats, etc., green, and about eighteen inches high, at the end of August. Last year, and in 1875, the rainfall during the autumn was so great throughout that the harvests were gathered with the greatest difficulty; and little or no autumn preparation of the land was possible. The years 1864, 1869, 1871, 1872, 1873, and 1876 were over large tracts of country almost totally destitute of rain in spring. When a drought sets in the sky becomes a clear deep blue, the dew deposited at nights decreases gradually ; and finally for nights in succession there is no dew at all. A few faint clouds may flit across the heavens, but with the haze in the horizon, the agriculturist may await rain anxiously from day to day in vain. A consequence of such clear skies is the extreme heat during the day, and the cold at nights, the mercury occasionally sinking above sixty degrees overnight to within a few degrees of freezing.

It is now almost an accepted fact that the droughty

inclination of a country's climate is due in greater
part to the absence of forests.   In Russia this effect
is probably intensified by the soil.   The soil is black,
and has therefore a greater attraction than any other
for the sun's rays, which it absorbs.   Much more
moisture is thus evaporated, and, owing to the lack
of forests, is not retained, but carried away by the
winds, leaving the atmosphere drier than before.

Looking at Table V. (page 29), we see that the
total crops for every two years vary but slightly;
from which fact we might infer that every other year
can be expected to turn out favourably.   In support
of this I may mention the observations made by Mr.
Schliept (Manager of Count Doshkoff's estates at
Khvalinsk); he noticed that the peasants, elated by
the success of a good harvest, invariably prepared a
larger amount of land in hopes of again having like
success; but that they as invariably were mistaken,
because it was altogether exceptional to his thirty-
five years' experience that two favourable years should
follow each other in immediate succession, but that a
good year almost always followed direct on a bad
one.   That three successive seasons, such as those of
1871, 1872, and 1873, should be droughty is quite exceptional.   I heard heart-rending accounts of these
three "hungry" years, as they were aptly termed.
Eye-witnesses told me the country looked perfectly
black for want of herbage, the grain lay in the ground
for months without germinating; and when rain

happened to fall, it all sprouted simultaneously, only too late to be of any use. In the meantime, peasants were starving, with their beasts, the very straw from the roofs being pulled down to feed the latter; whilst many of the former satisfied the cravings of their hunger by a rude sort of bread made from the farinaceous seeds of the more hardy weeds.

Although the frequent droughts are great hindrances to agriculture, yet the greatest losses incurred are caused by quite another phenomenon, namely, the prevalence of the S. E. wind,—a hot wind that appears to blow well nigh annually with great regularity for a few days. It parches the soil, scorches the upper ears of the wheat, shrivelling up the grain; and it very often brings on premature ripeness. It is bad enough to have one's grain lying ungerminating in the soil for want of rain, but to have one's crops almost annihilated by a few days with hot wind is still more trying. I remember once in 1876, while riding out with a landed proprietor, and admiring some crops of wheat and linseed which were expected to turn out very fair. When three days later I rode out again, it was very doubtful whether it would be worth while harvesting them at all.

# CHAPTER III.

## THE SOIL, AND ITS CAPABILITIES.

ONE of the most remarkable features of this country is the peculiar black colour of its soil. This soil, consisting of a crust of black vegetable mould one to three feet thick, extends from Bessarabia, in a north-easterly direction, all the way to Siberia; and although soil of a somewhat similar nature is found in the north of Germany, yet it is almost wholly peculiar to Russia. Its formation was long the subject of dispute amongst scientific men. At one time this soil was supposed to have been of a peaty nature; at another time it was believed that the subsoil on which it rests must have been covered by the ocean; but the absence of peat flora, and pebble-stones, together with other due characteristics, served to explode these views, and it is only lately that its formation has been definitely explained—originally by Huot, afterwards by Roobrekht.

The Steppes were originally covered with herbage, which, so long as untouched by man or his herds, died off on the self-same spot whereon it grew. Exposure to the atmosphere decomposed it, and it was in part transformed into a fertiliser, which was carried down into the soil by the rain or the water of the dissolving snows in spring. In proportion to the quantity of herbaceous parts thus carried down did the soil become of a deeper or lighter black. In these parts the vegetable mould lies on a light clay substratum many feet thick.[1] In one or two places in the south I have seen blue clay, and, where the land was much cut up by the rivers, also limestone. By observations taken from Tartar barrows, probably erected in the middle ages, and which possess now a slight layer of vegetable mould, Professor Karamzin calculates that 2400–4000 years are required to produce 2–5 feet of black soil. According to Klauss these black soils are characterised by the presence of *Stipa*, a large number of *Compositæ*, *Labiatæ*, and the *Almonds*, and by the absence of coniferous trees with their accompanying peat vegetation. Although the Flora of a country may not strictly belong to notes on its Agriculture, the two are here so intimately connected, that I shall make a few remarks about it. The crops of both proprietor and peasant are usually stifled with weeds, partly because the land is never weeded, and partly because insufficient or no care is bestowed upon the winnowing and dressing of seed-corn,

When land that has been farmed on the Three Course system[2] by either peasant or proprietor for

---

[1] I was present once at the sinking of a well 30 feet deep; there was nothing but clay. I have not heard of any sinkings to a greater depth at present. [2] See page 24.

many years is thrown aside to rest, it brings forth
a new variety of wild plants each succeeding season,
or, in other words, a natural rotation of plants takes
place ; and this with slight modifications is to the
best of my belief the case throughout the black-soil
plains. From a few rough notes kindly placed at my
disposal by Mr. Izdebsky of Samara, I am able to give
with a few additions some of the principal plants com-
posing this natural rotation.

1st year.　Cirsium, Carduus, Centaurea, Cichorium,
　　　　　　Chenopodium, Datura, Rumex, Atriplex,
　　　　　　Amaranthus, Artemesia, Bunias, Onobry-
　　　　　　chis, Sinapis, Brassica, Sonchus.　With
　　　　　　the exception of Setaria, almost no Gra-
　　　　　　mineæ or Leguminosæ.'

2nd year.　A few of the former (Artemesia, Sinapis,
　　　　　　Cichorium, etc.), with Melilotus glauca
　　　　　　and alba, Trifolium montanum, rus-
　　　　　　trum and repens, and many Gramineæ :
　　　　　　Agrostis, Briza, Alopecurus, Dactylis,
　　　　　　Festuca, and others.

3rd year.　Gramineæ, chiefly with Artemesia absin-
　　　　　　thium, Triticum repens.

4th year.　Artemesia, chiefly with Stipa (pennata ?),
　　　　　　which here makes its first appearance.

5th year.　Stipa alone is left, together with a little
　　　　　　Artemesia and a few others.

Afterwards, for as long as the soil remains untouched by the plough, the Stipa holds sway, and positively prevents other plants from taking root. The rapidity with which the rotation is gone through depends very much upon the previous cropping, the cattle driven over the land, and atmospheric influences, which either hasten or retard it, although essentially it remains always the same. Thus, if the last crop grown was very weedy, the first year's rest will display a proportionate large variety of plants ; if the crop was very much overgrown with couch, the couch will preponderate in the first year's rest. Cattle being left to pick up a scanty subsistence will, through their dung, introduce specimens which may not have been in the field before ; and others they probably annihilate, leaving often nothing but Artemesia absinthium standing. Rain increases the quantity of plants ; drought decreases the quantity, etc., etc. Land that has been farmed on the Steppe-system[1] will not present half the variety ; and the Stipa, instead of appearing in the fourth or fifth year, will appear already on the second and even first year's rest ; and again, if the land was very much exhausted by overcropping, the Stipa will not appear until after the tenth or twelfth year's rest.[2]

---

[1] See p. 26.

[2] In a thickly-populated district, where land is in demand, the stranger will be told the Stipa does not appear before the tenth or twelfth year ; in a thinly-populated part, he will be told the Stipa appears in the second or third year.

The Stipa, therefore, is regarded as an indicator, *par excellence*, of good soil. There is, besides the Stipa, as marking the quality of the soil, a plant called Vostretz, which on the lighter black soils indicates comparative poverty. I believe it to be a Poa, but I have never seen it in flower, only on the Steppes in autumn, when its grass was very thick. The people look upon it as an indicator of a poor soil, because it generally precedes Stipa.[1]

Besides the plants enumerated above, there are many others, such as Verbascum,[2] Asparagus, Salvia, Senecio, Vicia, etc., etc., which appear at irregular intervals until the stipa finally ousts all.

---

[1] The Dwarf Almond is another popular indicator of rich soil, since it appears only on land covered with Stipa, and which has not felt the plough for many years.

[2] This plant, containing much oleaginous matter, was until lately in great request amongst the peasantry to serve in lieu of candles. In early summer the eye is pleasantly attracted by its large yellow flowers studding the otherwise often lifeless pastures. It appears to have been imported as a weed by cultivation.

## TABLE I.

### Composition of Russian Black Soils from the Village Timashevo, Province of Samara.

**A.—**_Mechanical Analyses of 6 Soils (dried at 212° F.), made by Dr. Voelcker, 24th Nov. 1875._

| | Surface 1 | Subsoil 1. | Surface 2. | Subsoil 2. | Surface 3. | Subsoil 3. |
|---|---|---|---|---|---|---|
| Organic Matter and Water of Combination. | 10·09 | 3·79 | 15·70 | 5·30 | 10·76 | 4·95 |
| Carbonate of Lime .......... | 6·34 | 18·27 | 3·54 | 20·55 | 3·30 | 13·35 |
| Clay .......... | 51·55 | 39·74 | 51·15 | 38·65 | 54·13 | 63·25 |
| Sand .......... | 32·02 | 38·20 | 29·61 | 35·50 | 31·81 | 18·45 |
| | 100·00 | 100·00 | 100·00 | 100·00 | 100·00 | 100·00 |

| | Surface 4. | Subsoil 4. | Surface 5. | Subsoil 5. | Surface 6. | Subsoil 6. |
|---|---|---|---|---|---|---|
| Organic Matter and Water of Combination. | 10·16 | 6·97 | 11·50 | 3·70 | 9·27 | 7·30 |
| Carbonate of Lime .......... | 5·13 | 13·97 | 1·95 | 12·91 | 2·12 | 15·75 |
| Clay .......... | 58·31 | 54·36 | 53·06 | 48·21 | 48·51 | 52·08 |
| Sand .......... | 26·40 | 24·70 | 33·49 | 35·18 | 40·10 | 24·87 |
| | 100·00 | 100·00 | 100·00 | 100·00 | 100·00 | 100·00 |

## TABLE II.

### COMPOSITION OF RUSSIAN BLACK SOILS FROM THE VILLAGE TIMASHEVO, PROVINCE OF SAMARA.

B.—*Detailed Analyses of 6 Surface Soils (dried at 212° F.), made by Dr. Voelcker, 24th Nov. 1875.*

| | Soil No. 1. | Soil No. 2. | Soil No. 3. | Soil No. 4. | Soil No. 5. | Soil No. 6. |
|---|---|---|---|---|---|---|
| Organic Matter[1] and Water of Combination | 10·095 | 15·705 | 10·766 | 10·166 | 11·501 | 9·266 |
| Oxides of Iron | 5·052 | 6·407 | 4·501 | 6·666 | 5·352 | 3·746 |
| Alumina | 3·901 | 4·150 | 4·676 | 5·401 | 6·514 | 2·802 |
| Lime | 3·553 | 1·984 | 1·848 | 2·872 | 1·092 | 1·193 |
| Magnesia | 1·154 | 0·750 | 0·966 | 1·033 | 0·750 | 0·966 |
| Potash | 0·922 | 0·482 | 0·656 | 0·951 | 0·421 | 0·378 |
| Soda | 0·181 | 0·047 | 0·037 | 0·021 | 0·091 | 0·079 |
| Sulphuric Acid | 0·071 | 0·089 | 0·087 | 0·261 | 0·103 | 0·074 |
| Nitric Acid | 0·002 | 0·002 | 0·002 | 0·002 | 0·002 | 0·001 |
| Chlorine | 0·009 | 0·006 | 0·003 | 0·005 | 0·003 | 0·006 |
| Phosphoric Acid | 0·201 | 0·243 | 0·192 | 0·192 | 0·141 | 0·128 |
| Insoluble Silicates and Sand | 73·201 | 69·550 | 74·966 | 71·136 | 73·550 | 80·165 |
| Carbonic Acid, and loss in Analysis | 1·658 | 0·585 | 1·300 | 1·294 | 0·480 | 1·196 |
| | 100·000 | 100·000 | 100·000 | 100·000 | 100·000 | 100·000 |
| [1] Containing Nitrogen | 0·362 | 0·504 | 0·292 | 0·251 | 0·401 | 0·272 |
| Equal to Ammonia | 0·437 | 0·612 | 0·354 | 0·305 | 0·487 | 0·330 |

# TABLE III.

## COMPOSITION OF RUSSIAN BLACK SOILS FROM THE VILLAGE OF TIMASHEVO, PROVINCE OF SAMARA.

### C.—*Partial Analyses of 6 Subsoils (dried at 212° F.), made by Dr. Voelcker, Nov. 24, 1875.*

| | Subsoil 1. | Subsoil 2. | Subsoil 3. | Subsoil 4. | Subsoil 5. | Subsoil 6. |
|---|---|---|---|---|---|---|
| Organic Matter and Water of Combination | 3·79 | 5·30 | 4·95 | 6·97 | 3·70 | 7·30 |
| Oxides of Iron and Alumina | 4·79 | 7·65 | 9·90 | 12·50 | 9·75 | 9·75 |
| Carbonate of Lime | 18·27 | 20·55 | 13·35 | 13·97 | 12·91 | 15·75 |
| Insoluble Siliceous Matter (insoluble in Hydrochloric Acid) | 73·15 | 66·50 | 71·80 | 66·56 | 73·64 | 67·20 |
| | 100·00 | 100·00 | 100,00 | 100·00 | 100·00 | 100·00 |

### D.—*The Insoluble Silicates and Sand of Surface Soil No. 1. (73·201) consist of—*

| | |
|---|---|
| Oxide of Iron | 2·41 |
| Alumina | 6·66 |
| Lime | 0·85 |
| Magnesia | 0·98 |
| Potash | 1·26 |
| Soda | 0·52 |
| Silica | 60·52 |
| | 73·20 |

# TABLE IV.

## Composition of Russian Black Soils from the Village Timashevo, Province of Samara.

*Notes made by Mr. Izdebsky to the 6 Soils Analysed by Dr. Voelcker.*

| No. 1. | No. 2. | No. 3. | No. 4. | No. 5. | No. 6. |
|---|---|---|---|---|---|
| 31 inches black earth, below which is a mixture of clay and black earth 10 inches deep. Below this again is the clayey subsoil, the depth of which has been ascertained to within 9¾ feet. This grew rye this year (1875), producing a crop of about 12 cwt. per acre (?); it had been a bare fallow for the four previous years. | 26 inches black earth, then 8 inches mixture, below which clay subsoil, as in No. 1. This grew linseed this year (1875) : it had not been cultivated for several years previously. | 24 inches black earth, then 16 inches mixture, below which clay subsoil ; in the pit, which was 9¾ feet deep, water was reached. This had been pasture, untouched by the plough for 5 to 6 years. | 24 inches black earth, then 10 inches mixture, and below a subsoil of very firm clay. This year (1875) it grew rye ; before this it had 2 years' rest after a wheat crop. There is much Triticum Repens here. | 30–32 inches black earth, then 16 inches mixture, below this clay subsoil. This year (1875) it was fallow, previously wheat, before which rye, and before that pasture. | This comes from another estate. 24 inches black earth, then 10 inches mixture, below which the clay subsoil. This has been pasture for several years. |

Russians have a singular optimist opinion as to the capabilities of their black soil ; many believe there is no soil in the world to equal it. They all maintain that the blacker it is the richer it is, and look upon it as especially suitable for raising grain crops. Later on we shall see how far this high opinion is justified.

The land is easily worked. When it has lain at rest for many years, being used as pasture in the meanwhile, it becomes very firm, and requires six pair of oxen[1] to plough $1\frac{1}{3}$ acres in a day, with a Ransome's R. N. F. W. plough. On exposure to the atmosphere, it quickly slakes down, and becomes very friable, the roots only of the rough herbage making it work hard. The following year two pair of oxen can easily plough the same quantity in the same time, with a Ransome's H. C. plough. Even in the first year a tolerably satisfactory tilth can be obtained. It absorbs moisture readily, and in consequence of its dark colour is soon dried. It can be worked when wet without difficulty, as it cakes but slightly. Grain sown in the wet soil comes up far healthier than when sown in dry soil.

Cultivation is carried on in several ways. While in the Cis-Volgian provinces clover has already settled down permanently in the rotation, and farm-yard manure now rarely pollutes the rivers, the provinces

---

[1] Two to three pair of Sussex oxen.

on the left bank know of clover but by name, and manuring is still an exceptional phenomenon. Of the methods in which land is cultivated two are systematic, and one unsystematic. Unsystematic cultivation consists in cropping at irregular intervals, without due regard to rest or to a rotation. A good deal of land is cropped in this way, being chiefly such land as is let for a year or two to the peasants, or any one willing to hire; after which it lies at rest awaiting a new hirer; being relet to the first comer who offers a reasonable rent, no matter what the last crop was, or how short the rest it has had in the meanwhile. Such land is generally excessively weedy. To get rid of the tall weeds that spring up beween the croppings, firing is often resorted to in spring, during which time the whole country occasionally appears to be in a blaze.

The Three-Course is the system chiefly in vogue where the population is comparatively thick, and where the land is much exhausted from previous cropping, being in consequence the sort of land to which farm-yard manure is first applied. All the peasantry pursue this system, and most of the larger proprietors follow it on at least a part of their estates. The Three-Course system consists of a winter crop, after one year's rest, succeeded by a spring crop, thus:

1. Rest one year.
2. Winter wheat or rye, chiefly the latter.

3. Spring wheat, barley, oats, millet, and buckwheat, or peas.

The preparation is not everywhere the same, there being slight deviations in practice according to locality. It may, however, be summed up as follows : The bare lying land is ploughed up and harrowed towards the end of June. A fortnight later it is ploughed and harrowed again for the sake of destroying any weeds that may have sprung up in the meantime. About mid-August the seed is sown broadcast, and ploughed in with the sokha.[1] The sokha leaves the land in a rough condition, which is, however, said to preserve the young plant, either by preventing the wind from blowing away the snow, and exposing the plant to the cold, or by keeping the water in the furrows in early spring, and thus aiding the young plant to take a firm hold should the season turn out at all droughty. In consequence of autumnal rains, the young rye often pushes too rapidly forward ; this is overcome by eating it down before the winter sets in. The winter crop having been removed, the land is ploughed up with the sokha (a very poor performance, as the implement barely goes three inches deep) ; and the following spring the seed is sown as soon as it is possible to get on to the land, and while it is still wet with the waters of the dissolving snows. The grain comes up all the better for being sown in the

---

[1] See page 43.

wet soil, and has the advantage of a fair start in case of drought.

The Long-Rest or Steppe-system consists in ploughing up land that has been at rest and pastured from 5–15 and more years, and sowing it 2–3 years consecutively with spring crops. This system is to be met with on all the larger estates, and almost exclusively on the more distant land from the east to the south of Samara. It is the system *par excellence* under which the Byaylatoorka (White Turkish) wheat flourishes most successfully. The number of years the land rests varies in the north from 12–15 years, and towards the south and east, 5–10 years. By the time the land has lain, say six years, Stipa (Covil, as it is popularly called) would be the only remaining grass, a grass scanty enough at all times. It is ploughed up in the summer, and sown in earliest spring with Byaylatoorka; sometimes bearing the same three years in succession, and sometimes only two years. Lately linseed has been grown largely ; and a few years ago, oats were introduced as crops to succeed a one year's growth of Byaylatoorka. Winter rye occasionally follows a spring wheat; it is in such cases sown broadcast by the hand, and harrowed in on the *un*broken wheat stubble.

The crops obtained under the first-named systems are, as may be expected, very meagre. The data I have concerning them are not thoroughly reliable, as they were not extracted from regularly-kept books, but given me at random out of the heads of my informers. From the various information thus received, I should calculate the harvests under the Three-Course system as follows :

|  | POODS. | | | LBS. | | |
|---|---|---|---|---|---|---|
| Rye ............... | 120 per sotelnaya dessyatin [1]= 960 per Eng. acre. | | | | | |
| Oats ............. | 100 | ,, | ,, | = 800 | ,, | ,, |
| Spring Wheat... | 60 | ,, | ,, | = 480 | ,, | ,, |

The above figures are averages extending over many years, and do not speak well either for capabilities of soil or cultivation.

I was rather more fortunate in getting statistics relative to the crops obtained with the Long-Rest system. It is not only very rare that farm-books are kept, but still more so is it to find them combined with meteorological notes. In Table VI. given below (see page 32), we have the harvests for the last ten years (1868 to 1877 inclusive), on the estates of a merchant, Mr. Plyeshanoff, at Ivanoffsky Khooter, in Nikolayeff district, about 150 versts (100 miles) south-east of Samara. The soil there is not quite so black as those which were analysed by Dr. Voelcker, but nevertheless the soil is considered a very rich one. Mr. Plyeshanoff's land lies bare-fallow (as pasture) six years. Byaylatoorka wheat,

---

[1] For table of weights and measures, see end of Work.

linseed, and oats are the crops then successively sown,
after which it is again left fallow. The soil thus
gives three crops successively once in nine years. In
Table V. we have the harvests grown on the estates
of Mr. Ryagozhin, at about 60 versts (40 miles)
from the former. The soil here is very black, and,
so far as the knowledge of the proprietor goes, was a
virgin soil when first ploughed up by his father in
1846. The land is ploughed up once in six years,
sown with Byaylatoorka wheat, and then left again to
rest. It thus gives one crop every seventh year.
Lately, however, linseed has been introduced to
succeed the wheat. This second table is the more
interesting because of the long range of years over
which it extends, and enables one to form a more
accurate estimate of the capabilities of the soil. The
averages for wheat are 564 and 499 lbs. per acre
respectively. These averages calculated from the
two tables must be considered above the mean for
the whole country, for the soil is, according to Russian
opinion, first class, the best implements are in use,
and the best possible cultivation pursued. The
ploughing was in both cases 7–8 inches deep, and
the land was freer from weeds than is usually the
case.

But in order to judge of the capabilities of the soil,
a crop must be taken which, besides being grown
under the best Russian knowledge, had also the
advantage of a favourable season. The largest crop

TABLE V.—*Statement of Mr. Ryagozhin's Wheat Crops for 32 years, 1846–1877 inclusive.*

| Year | Total in lbs. per acre. | Biennial Totals in lbs. per acre. | Biennial Averages in lbs. per acre. | Unfavourable Atmospheric Influences. | Remarks. |
|---|---|---|---|---|---|
| | | | | FIRST DECADE. | |
| 1846 | 376 } | 856 | 428 | | The irregular quantities harvested during the first decade depended, probably, as much upon the bad preparation of the soil as upon a bad selection of soils, one year a superior piece being chosen, and the next year an inferior one. |
| 1847 | 480 } | | | | |
| 1848 | 174 } | 922 | 461 | | |
| 1849 | 748 } | | | | |
| 1850 | 376 } | 838 | 419 | | |
| 1851 | 462 } | | | | |
| 1852 | 768 } | 1300 | 650 | | |
| 1853 | 532 } | | | | |
| 1854 | 572 } | 914 | 457 | | |
| 1855 | 342 } | | | | |
| | | | | SECOND DECADE. | |
| 1856 | 582 } | 968 | 484 | | In the first 8 years of this decade, in consequence of more regular cropping, and a more careful preparation of the soil with the usual native implements, the biennial crop totals |
| 1857 | 386 } | | | | |
| 1858 | 818 } | 972 | 486 | | |
| 1859 | 154 } | | | | |
| 1860 | 304 } | 1060 | 530 | In 1860, a S. E. wind burnt many of the ears, causing great emptiness amongst them. | |
| 1861 | 756 } | | | | |

*Statement of Wheat Crops—continued.*

| Year | Total in lbs. per acre. | Biennial Totals in lbs. per acre. | Biennial Averages in lbs. per acre. | Unfavourable Atmospheric Influences. | Remarks. |
|---|---|---|---|---|---|
| 1862 1863 | 448 512 | 960 | 480 | In 1862, microscopic insects annihilated much grain. In 1863, the S. E. wind again did much damage. | were remarkably uniform —from 960–1060 lbs. per acre; but in the last two years of this decade the exceptional drought of 1864, and the at first totally unsuccessful sowing of 1865 (from which it appeared that the seed was bad—up to 120. lbs. per acre having been sown), the last biennial total reached only 836 lbs. per acre. |
| 1864 1865 | 196 640 | 836 | 418 | In 1864, there was an exceptional drought ; the S. E. wind appeared again with great severity, burning and causing the upper ears to be perfectly empty. | |
| | | | | THIRD DECADE. | |
| 1866 1867 | 540 504 | 1044 | 522 | In 1866, locusts, hail, and slight S. E. wind affected the crop. In 1867, a hot wind did much damage. | At this time, iron ploughs, heavy harrows, steam threshers, and other improved machinery first came into use ; the biennial totals began to increase remarkably. |
| 1868 1869 | 1040 200 | 1240 | 620 | In 1869, drought in June, grain-insects, and lastly the ears were harmed by a hot wind. | |

| Years | | | | Notes |
|---|---|---|---|---|
| 1870<br>1871 | 1190<br>256 | 1446 | 723 | In 1871, drought during the whole summer. The wheat was burnt up at the roots, so that many ears were perfectly grainless. |
| 1872<br>1873 | 114<br>92 | 206 | 103 | In 1872, the spring was droughty; rain fell towards end of summer, so that land was overgrown with weeds. The year 1873 was droughty in spring and summer. |
| 1874<br>1875 | 761<br>542 | 1303 | 651½ | In 1874, the autumn was very wet and unfavourable for work, so that much grain was lost. In 1875, the grain was of a bad quality, owing to S. E. winds at a high temperature. |

## FOURTH DECADE.

| Years | | | | Notes |
|---|---|---|---|---|
| 1876<br>1877 | 230<br>880 | 1110 | 555 | The spring, summer, and autumn of 1876 were rainless. On July 4, 1877, the heat was terrible, causing much grain to ripen prematurely. |

Average per annum for 32 years—1846-1877—amounts to 499 lbs. Eng. per acre.

## TABLE VI.

*Statement of Mr. Plyeshanoff's Crops for the last 10 years.*

| | Wheat, lbs. per acre. | Linseed, lbs. per acre. | Oats, lbs. per acre. |
|---|---|---|---|
| 1868 | 1000 | ...... | ..... |
| 1869 | 104 | 336 | ..... |
| 1870 | 1000 | 1600 | ..... |
| 1871 | 184 | 40 | ..... |
| 1872 | 224 | 120 | 200 |
| 1873 | 264 | 224 | ..... |
| 1874 | 880 | 264 | 744 |
| 1875 | 680 | 600 | 2240 |
| 1876 | 264 | 96 | 760 |
| 1877 | 1040 | 320 | 2000 |

Average for 10 years' growth of wheat ........................... 564 Eng. lbs. per acre.

,, ,, 9 ,, ,, linseed .................. 400 ,, ,,

,, ,, 5 ,, ,, oats ..................... 1189 ,, ,,

recorded under these conditions was in 1870, when 1190 lbs. English (20 bushels) per acre (Table V.) were harvested—a result an English farmer would justly grumble at. Therefore the cause of the small produce obtained must be looked for elsewhere than in unfavourable climatic influences. In studying the analyses made by Dr. Voelcker, it will be found that the soil has a very fair amount of every element of fertility. The predominance of the organic matter was explained when the formation of the soil was treated of, and it is probably due to this excess that the soil gives such small crops. When the land is left at rest, the Stipa ousts all other wild plants; while, at the same time, the Stipa itself becomes thinner and poorer; and when the ground has lain thus many years, the Stipa so thinly covers the soil that the ground becomes speckled with small black patches, upon which nothing is growing.[1] The cause no doubt lies in the want of porosity, induced by the large amount of vegetable matter, which prevents the penetrative and chemical action of the atmosphere. In the same way when ploughed up for one crop only, the atmosphere has not sufficient time to work, and the crop is poor. This may be seen by the better crops of linseed and oats obtained in the second and third year of cultivation at Mr. Plyesha-noff's, and by the fact that a second growth of wheat

---

[1] This is not to be confounded with effloresence of salts. See page 37.

D

is always better than the one it follows, provided the climatic conditions remain the same. When Dr. Voelcker analysed the samples he recommended liming, I believe, in order to get rid of this super-abundance of vegetable matter.

The Russians like to plough up large broad slices, only 4–5 inches deep ; paring being unknown, the fibrous roots of the rough herbage hold the soil firmly together, and prevent the penetration of the atmosphere. When the time comes for harrowing, the teeth cannot penetrate, and the soil is left in a very superficial state of preparation. The teeth of the harrows, being always askew, or out of position, repeated harrowings are necessary, and the work performed is abominable. Drainage is not required, the water trickling quickly through the soil, which dries up remarkably fast.

Several landed proprietors have made attempts at farm-yard manuring, but few successfully ; while some few maintain that the straw of the farm-yard manure does harm by increasing the friability of an already friable soil ; others decry it as causing the crops to be laid by over-luxuriance. Both effects are not at all unlikely. In other parts of Russia, farm-yard manure on black soil was introduced long ago, and was probably successful because the land had been pretty well aired before its introduction ; otherwise, I am at a loss to understand why it should not meet with success in Samara province. Very likely landed proprietors find that manure does improve the crops, but that the increase obtained does not cover the increased expense for labour. Mr. Boolighin, at Krotoftka (and his father also before him) has been in the habit of farm-yard manuring. He considers it pays him, but is not quite satisfied with the results.

Two gentlemen (Mr. Shiskoff, of Arkhangelsky-Pyenyoffka, in Stavropol district, and Mr. Tishinsky, of Kinagyevo, in Tamboff province) were successful with their attempts at fertilizing with super-phosphate of lime; but the increased weight of crops was not considered sufficiently remunerative to warrant a second trial.

In order to find out whether it was possible to plough deeply without bringing up any sour subsoil, a small piece of land at Timashevo was tested by digging up two feet of subsoil, turning it over, and growing wheat. Owing to some stray peasant cattle, however, the weight of the crop could not be ascertained; but it was considered a very heavy one by all who saw it. Under the three-field system, the poor crops can only be ascribed to sheer bad cultivation (the sokha doing the greater part of the work here, and working only 3–4 inches deep), and to the foulness of the land, the corn being usually stifled with weeds.

The quantity of weeds to be met with in all the fields, and the complacency with which they are tolerated, continues to be a great drain on the soil, to the detriment of the crops. Many agriculturists will deny the presence of weeds in their fields, but whenever I went to look, I found them simply choking the corn. Bindweed, corn-cockle, charlock, twitch, docks, thistles, common sow-thistle, wild mustard, rape, artemesia, atriplex, amaranthus, and a host of others, are only too common. Dodder in flax is most harmful.

The foul state of the land can hardly be wondered at, seeing that little or nothing is done to remedy the evil. An attempt is made to exterminate the twitch. During the hot days shortly before sowing, the sokha is sent through the land ; an operation which certainly destroys some of the twitch, but which, at the same time, so thoroughly dries the soil that the seed when put in will lie for weeks without germinating. In mowing or reaping, the peasant is too lazy to cut down the stronger and thicker weeds, such as thistles, artemesia, verbascum, etc. : these run to seed, and are thus eternally procreated and propagated. Gross carelessness is also manifested in the dirtiness of the seed sown, the seed-corn being usually very badly dressed.[1] The peasants are the greatest sinners in this respect, and their land is also the foulest. To the south and east, the land is much cleaner, probably because it has not been so long under cultivation.[2]

---

[1] The dressing is usually performed by tossing the grain up in the air, whereby the wind catches and blows away the lighter weed-seeds and chaff.

[2] Smut in the grain, and more especially ergot in rye, are exceedingly common.

Grain, and more especially linseed, comes to England full of weed-seeds. Here is an explanation. At Timashevo last year some well-dressed linseed was sold in town at 1·40 roobles per pood ; the weed-seeds extracted by the cleaning and dressing, consisting of all the worst annuals that grow here, were sold at seventy-five kopecks per pood, the merchants having applied specially for them in order to remix with the better-dressed linseed and wheat.

The great pests of the animal world are the Sooslicks, which are said to do great damage to the crops.

Efflorescence of salts are occasional hindrances to agriculture. Where these salts appear nothing will grow ; happily they are restricted to the south-east of the provinces bordering on the sandy deserts.

The amount of seed-corn sown varies greatly. Wheat is sown at the rate of 40–80 lbs. per acre, average 65 lbs. ; rye and oats 60–110 lbs. per acre, average 90 lbs. ; barley usually 90 lbs. ; and linseed 24–36 lbs. per acre.

# CHAPTER IV.

## OTHER AGRICULTURAL PRODUCTS.

THE other agricultural products which are sown in smaller quantities are sunflower and poppy for oil, rape, and mustard. Sunflower is not a paying crop unless great care and watchfulnesss be bestowed upon it. The peasants are exceedingly fond of the seed, and will pilfer half the crop before it is ripe.

Melons and water-melons are grown largely by the peasantry, who, in common with all Russians, are particularly fond of them. They are grown best on newly broken-up pasture. Cucumbers are also cultivated on a large scale.

In the gardens of the peasantry (which, by the bye, are as foul as their fields) are grown hemp and flax, which are spun at home during the winter. In other parts of Russia, flax is grown in regular rotation for the fibre for export. In these parts, however, flax is grown for the seed (linseed) only,—the fibre being thrown away, or burnt with superfluous straw. In the kitchen-gardens of the peasantry we find also potatoes, cabbages, etc. : the latter are in request for the preparation of the famous Russian cabbage - soup ("shtchee").

Landed proprietors take great pride in the manage-

# TABLE VII.

## Composition of Sugar-beets grown at Timashevo, Samara, Russia, 1876.

*Analyses made by Messrs. Mandyeleff and Pavloff.*

| Variety of Beet. | First Collection. | | Second Collection. | |
|---|---|---|---|---|
| | Weight in Russ. lbs. | Percentage of Sugar. | Weight in Russ. lbs. | Percentage of Sugar. |
| Collets Verts (seed from Vilmorin) | 1 5/8 | 9·77 | 2 1/4 | 12·02 |
| „ „ „ | 1 4/5 | 10·16 | 1 1/2 | 12·50 |
| „ „ „ | 1 1/4 | 10·87 | 1 | 13·56 |
| „ „ „ | 1 3/16 | 11·07 | ...... | ...... |
| Collets Roses „ | 2 | 4·26 | ...... | ...... |
| „ „ „ | 1 1/11 | 7·38 | ...... | ...... |
| „ „ „ | 8 | 7·69 | ...... | ...... |
| Imperiales acclimatées „ | ...... | ...... | 2 1/4 | 8·93 |
| „ „ „ | ...... | ...... | 1 1/2 | 10·31 |
| „ „ „ | ...... | ...... | 1 | 11·96 |
| Vilmorins ameliorées „ | 4 | 6·89 | 2 1/4 | 9·63 |
| „ „ „ | 1 3/4 | 7·37 | 2 | 10·28 |
| „ „ „ | 1 7/16 | 7·90 | 1 1/8 | 11·13 |
| „ „ „ | 1 | 9·68 | 2 1/4 | 13·43 |
| Imperials (seed from Magdeburg) | 2 1/8 | 8·69 | 2 | 8·64 |
| „ „ „ | 1 1/2 | 11·04 | 1 3/4 | 10·20 |
| „ „ „ | 1 5/8 | 11·76 | 1 3/4 | 10·90 |
| „ „ „ | 1 7/8 | 13·91 | 4 | 12·12 |

The second collection took place about three weeks after the first.

ment of their flower and kitchen-gardens, orchards, and hot-houses, which are quite as productive as those of their *confrères* in England.

While at Timashevo, two new products were intro-duced—one was sugar-beet, with a view to the manu-facture of sugar ; the other was tobacco. The foreign colonists from Khvalinsk, downwards along the Volga, have long grown tobacco ; but it was never attempted so far north as Samara before. So far as two years' experience goes, I feel convinced that tobacco may be grown successfully ; but as to sugar-beets I still en-tertain considerable doubts, on account of climatic conditions. The year 1876 was exceptionally pro-pitious for sugar-beet, and the analyses on Table VII of sugar-beets grown that year, made by Messrs. Man-dyeleff and Pavloff, will give an idea of their qualities. The autumnal rains of last year caused too great a luxuriance among the beets. At first, when the proper time arrived for harvesting, it was too wet to collect them ; and later, they grew to such large dimensions, that it was not considered worth while to have them analysed. The objection I found against their cul-ture was the short duration of the summers, which barely allows them to arrive at the proper maturity.

Mr. Rickard at Orenburg has been very successful in acclimatising English Black Tartarian oats.

# CHAPTER V.

## FODDER.

As already mentioned, clover has settled down as a regular crop in the Cis-Volgian provinces. In the Trans-Volgian, however, with one or two exceptions, clover is unknown, agriculturists relying upon the propitiousness of Providence for their fodder supplies. Hay is chiefly prepared from the rank grass that grows luxuriantly enough on the low-lying lands, river banks, or other hollows which, owing to their position, cannot be cultivated, being flooded in spring, when the snows dissolve, and rivers are swollen, but which dry up later in the summer. There is also always more or less land on the steppe or long-rest system which already on the third year's rest gives good hay, consisting, for the most part, of couch grass, afterwards ousted by the Covil or Stipa (pennata ?) Covil forms a highly-nutritive, but at the same time, an excessively coarse hay, on account of the great quantity of siliceous matter it contains. The hay harvested from Covil meadows is pitifully small in quantity. I cannot give the exact amount of Covil hay harvested, as no man I came across had felt himself sufficiently interested to have a crop weighed ; but I should say a good crop might give sixty poods per shestdaysyattaya dessiatin, or about five cwt. per acre.

Since land cultivated on the steppe-system does not grow sufficient natural herbage for the sustenance of stock until the second or third year's rest, much capital lies for a time uselessly locked up. To remedy this, a spirited Cossack gentleman, Mr. Obratnoff,[1] attempted the introduction of artificial grasses. After the removal of the crops he sowed the following varieties : sanfoin, lucerne, Timothy-grass, Swedish clover, and German panic-grass. This attempt was made three to four years ago. The two first-named grasses throve, and still thrive ; but it is questionable whether this trial is a success from the chief, *i. e.*, the economic point of view, for which it was instituted.

---

## CHAPTER VI.

### IMPLEMENTS AND MACHINES.

THE native implements are one and all most primitive and bad ; the sooner they are totally supplanted by improved foreign machines and implements the better it will be for Russian agriculture.

---

[1] This gentleman obtained prize medals at the Philadelphia Exhibition for his wheat ; Mr. Plyeshanoff, previously mentioned, obtained medals at the same Exhibition for his linseed.

There is, however, one native implement—the sokha—which may be interesting on account of its peculiar construction. It consists of a light square frame, the two sides of which form the shafts; two curved wooden tynes inclining towards one another project below, each one being tipped with a flat iron share; the two shares together have the form of a ridging plough-share. The tynes are further securely fixed in position to the shafts by some strong twine. A slightly convex trowel-shaped small mould-board (if so it can be called) is lightly attached close above the shares; this mould-board has the motion of that of a turn-wrest plough, and the implement is used as such. It is not intended for ploughing arable land to a greater depth than three to four inches, and one horse suffices to draw it easily. The implement might be described as a one-way single-tyned cultivator.

The introduction of improved agricultural implements is progressing. Messrs. Ransomes, Sims, and Head, the well-known agricultural engineers, are the great importers of all implements and machinery into Russia. A favourite plough in this part of the country is their wood-beam plough, II. B. series. Where piece-work is in vogue the light multiple ploughs, M. E. D. M., are preferable, as with the former the labourer is enabled to scamp his work by steadying it at a slight inclination. Moreover, the multiple ploughs have well-known economic advantages, which need no repetitions here. For breaking up pasture, the iron heavier ploughs of the R. N. F. W. type, both single and double furrow, are most suitable, and are to be seen frequently in practice on the larger estates.

Eckert's Disc Broadcast Sowing-machine is very popular on the Steppes. Like many other German machines, it is copied from the English, being in fact an exact, but weak, imitation of the one first patented and manufactured by Messrs. Ben. Reid & Co., of Aberdeen. This machine is remarkably adapted for agriculture in Russia, where the size of the fields is unlimited, varying often from half a verst to a mile in length. The machine is broad, exceedingly simple, and so light of draught that one Russian horse suffices to sow from 25–30 acres in one day. Should drills ever come into use here, it may be safely predicted that the American drills, such as the "Farmer's Favourite," etc., will compete successfully against any of the present English machines; for cheapness and lightness of draught the former are not to be equalled. The shape of their coulter is also peculiarly adapted to the black soil.

Steam threshing-machines have in many localities become indispensable. This is the case in the great corn-growing districts of Novo-Oozensk, and Niko-layeff. As previously mentioned, it is of the utmost importance to get the grain threshed out immediately after harvest, an end steam threshers serve best to attain. At present there are very few large estates which do not possess a steam thresher. In the neighbourhood of the town of Samara, a merchant, Mr. Shabayeff, lets out steam threshing-machines. He possesses three ; and although his charges are high, he finds full and lucrative employment for all. Shortly before leaving Russia, I heard of several applications made by the peasants for the loan of steam threshers of their respective neighbouring landed proprietors. This, if true, is another step in the right direction.

Much grain is still threshed out by the flail, and trodden out by horses ; the latter mode is chiefly confined to the threshing of White Turkish wheat, which is very tough.

Reaping and mowing machines have come into universal use. Wood and Samuelson are well represented ; but the Johnstone Harvester Company has probably sent more machines to Russia than all the other manufacturers put together. The use of mowing machines will no doubt tend to the eradication of those big annual weeds which the peasant refuses to cut down either with scythe or sickle. Many may think that the great advantage arising from the use of this class of machinery lies in the cheapened production. This is, however, not always so. In droughty seasons, like 1876, the peasants reap and mow decidedly cheaper than the machines. In Nikolayeff and Novo-Oozensk (where 7½-25 roobles per sotelnaya dessyatin, 4*s.* to 14*s.* per acre, are paid for reaping and binding), greater profits are to be realised by binding reapers for very often where the simple reaping machine is used the peasants charge all the more for binding and collecting.

As a rule, the peasantry take most kindly to new machinery and impléments ; and although opposition is manifested occasionally by the smashing of a drum, the withdrawal of important bolts, etc., such occurrences are more the result of individual malice than collective hostility.   I have always heard and found that the peasants delight in any employment with machines, no matter what sort, from a simple broadcast to a portable engine.   They learn the parts remarkably fast, and rise considerably in their own estimation, as well as in that of their fellow-labourers. Would I could say as much for the bailiffs !

---

## CHAPTER VII.

### AGRICULTURAL HORSES.

THE agricultural horses are the well-known Tartar ponies.  They are reared in large troops on the uncultivated Steppes in an almost wild state, being, with the exception of guards, left entirely to their own resources for sustenance.   In summer they graze, and in winter kick away the deep snow to nibble at the poor herbage underneath.   After droughty summers, or when frost has set in before the snows have fallen, the poor horses fare very badly; and notwithstanding their thick, shaggy skins, they become terribly skinny. Few, however, succumb, for any animal that can withstand the first winter is enduring enough to sustain all future hardships. The same may be said of all live stock, and of peasants' children as well.

From four to six years the horses are in fine condition, and are then brought to market, and when sold have to be broken-in. Breaking-in takes three or four weeks, during which the poor beasts are daily securely harnessed to a cart, and made to go. An old horse leads; and if they do not move, they are lashed until they do. Being unaccustomed to hay and straw, they refuse at first to eat; they rapidly lose in strength, and are the less able to be violent. Their mouth soon becomes as hard as wood, and it is greatly to be wondered at that so few turn out vicious. This rough treatment makes it impossible to shoe them as we do in civilised countries. For this end, at every smith's will be seen a sort of small scaffolding, in which the animals are securely bound. The foot to be shod is then raised, and fixed in a convenient position for shoeing. This leads to careless shoeing, whereby more harm is done than good, and many owners in consequence prefer to let their horses go about unshod. The total absence of anything like a hard macadamised road exempts unshod horses from any inconvenience arising from a too rapid wearing away of the hoof ; but in wet weather and in winter the absence of shoes entails a considerable increase of labour on the horses.

There are several studs for town and carriage-horses, but that of Mr. Stobayoos is the only one of any value or importance. At this establishment, under the management of Messrs. Balk and Hiegel, are produced some of the finest thorough-bred carriage horses to be met with in the streets of St. Petersburg.

The price of agricultural horses depends intimately upon the harvests. A nag worth forty roobles (£5) after a bad harvest, may be worth double that amount next summer, in anticipation of a good harvest, and *vice versa*. The price of an agricultural horse varies from 35 to 100 roobles.

Old horses and cattle are sometimes said to be fattened with small doses of arsenic. Both Tartars and peasants have been pointed out to me as men who administer arsenic to their beasts ; but when taxed, they invariably denied doing so. Nevertheless, it has come so often under my notice that there must be some truth in it. Even women appear to administer it to themselves—female stoutness being considered a point of beauty among the peasants.

In the more immediate vicinity of the town of Samara, Bashkeer and Kirghiz mares fetch a higher price than ordinary horses, owing to the demand for mares' milk at the Koomis-Cure establishments. These institutions, at the head of which stands that of Dr.

Postnikoff, are peculiar to Samara province. Koomiss is fermented mares' milk, and is the national beverage of the Kirghiz and Bashkeers. Some twenty years ago Dr. Postnikoff discovered the beneficial effects this fermented mares' milk had on consumptive individuals, and remarked at the same time that the above-named hordes were totally free from all pulmonary affections. He subsequently introduced the koomiss as a cure for consumption ; and at present the innumerable consumptives who repair during the summer to Samara to partake of koomiss are sufficient evidences of its advantages. The climate of Samara is, moreover, considered the most favourable for this especial cure. The mares are found to give best milk when grazed on their native Stipa-covered steppes ; and it is said that, owing to the successive generations during which it has been customary to milk them, their teats develop to a greater extent than that of any other race. A good milking-mare will fetch £1–3 per head above the usual price.

---

## CHAPTER VIII.

### CATTLE.

OF horned cattle there is very little to be said, for agriculture has not advanced so far as to make cattle-breeding a lucrative investment. Bullocks are

E

chiefly employed for draught purposes; but even then, choice animals possessing good working quali- ties are not selected for breeding, but the bull is allowed the full run of all the cows without dis- tinction. They are of all colours. Silver-greys are often predominant, and present a greater uniformity amongst themselves than those of any other colour. The greatest diversity is manifest in the peasants' stock, which only agree in one point, namely, in size. I believe the poor food they have to put up with from an early age to be the main cause of their diminutive or stunted stature. Mr. Hiegel, at Al- exandroffka, showed me some very large draught oxen, which he assured me were the offspring of the peasants' ordinary stock, but which had been well fed and cared for during the first 2–3 years of their existence. It is, however, characteristic of Russian oxen to fill up quickly and well in a very short time.

In autumn the cattle are driven from their pastures into the open yards (adjoining every village home- stead, and surrounded by wattle-hedges or sheds on the most exposed quarter), where, unless work has to be performed, they remain the whole winter, feeding chiefly upon rye straw. The meagreness of such food, the great waste in the continuous liberation of heat, and the loss sustained by the excessive growth of hair all over their bodies required to support them against the effects of such a long exposure to the cold, all aid together to bring them down to the very lowest possible condition. Spring comes,—many through sheer weakness can barely crawl out of their yards ; they shed their hair, and then their thoroughly emaciated state is plainly visible. They look, in fact, no better than bags stuffed with bones. Shortly before being set to work they are fed upon a little hay, in addition to the straw ; and three to four weeks after the snows have melted they are already able to graze on the pastures. The sudden change from dry to succulent food appears to have no, or at most little, effect on their hardy constitutions.

A short rest after spring operations brings us to the middle of June. If we look at them then, they show a remarkable contrast to their previous miserable condition,—their skins have become soft and glossy, they have filled out, and are, to a certain extent, plump. One can then occasionally pick out a few very good specimens, possessing tolerable rectangular frames, which, under proper care and selection, might be made to produce some good fattening stock. Their price ranges from 20 to 35 roobles.

The native cows are ridiculously poor milkers ; but being badly housed and little attended to, much milk cannot be expected. From spring to autumn they are driven every morning after milking on to the pastures, where they remain until nightfall, when they return home, each individual animal finding its way to its own yard. They are thus usually milked at 4 a.m. and 8 p.m. If they were brought home earlier it would necessitate feeding in the yard with hay, which until after harvest is generally not to be had ; but I very much doubt whether sixteen hours interval between the two milkings is an economical arrangement. Price varies from 15 to 40 roobles.

Several energetic proprietors are introducing Dutch, Danish, Swiss, and Tyrolese bulls for improving the milk-giving qualities. Many who would most gladly import strangers for breeding are afraid to do so in consequence of the permanent exposure to the risks of the cattle-plague. The chief preventive to improved methods of breeding is the scanty population. A scanty population has few wants ; and so long as little meat is consumed, and the people are satisfied with the meagre crops obtainable from unmanured land, there is little prospect of any permanent amelioration in the live stock.

---

## CHAPTER IX.

### THE CATTLE-PLAGUE.

THE steppes of Russia are generally looked upon as the home of the cattle-plague. In all probability, however, this view is not quite correct, for the plague appears to be quite as much at home in the interior of Asia. It never dies out, but is fitfully at move in the same old tracts. In some places its recurrence is so frequent that cattle-keeping is quite out of the question. In others, again, it is almost unknown. This is explained by the fact that those villages which the plague visits so frequently are situated on the trunk-roads that form a part of the great system of internal communication of the country, and in such localities it is often of annual recurrence ; whereas, when it does attack the cattle in situations distant from these trunk-roads, its visit is anomalous, and after one visitation it rarely comes again. I met it near Kazan, on the Kazan-Orenburg road ; and I heard of it on the same road close to Orenburg, and then on the road to Oofa. Previously I had heard of it on the road between Samara and Orenburg, along the right bank of the Volga, on the Sizerayn Simbirsk road, lower down at Khvalinsk ; and then again in the province of Saratoff.

During the last two years it appears to have been very quiet, and only at Kazan did I meet it in full activity. The authorities were doing their best to prevent its spread ; but they were perfectly powerless, for the peasant is not easily dealt with, and to overcome his stubborn ignorance is a difficult task. The head Elder of every village has full instructions as to what is to be done; but unfortunately, even if his intentions be good, he does not generally possess sufficient ability to carry them out. A peasant thinks it bad enough to lose his ox without having to lose the hide; so that, unless narrowly watched, he will take the hide home, and there infect the rest of his own cattle and that of the community. Next to his horse (the Little-Russians possess oxen exclusively), the draught-ox is his most important possession. His ox does all his field work, takes him to the fields or forest, carries his grain to market, and brings back beams or other necessary commodities from the town. In price it is not high, and, above all, during the long winters straw is quite sufficient to keep it alive. If then, by the plague he loses so much, why should he at least not be allowed to sell the skin ? Explain to him that he will thereby infect other animals, fine him, imprison him, nevertheless, if he gets the opportunity, he will skin the beast, and bury it so slightly that the dogs scratch away the light layer of soil, feast on the carcase, and help, together with the crows, to carry the infection a step further. Bleached bones of cattle on the steppes are common enough, —the peasant generally lets his beast lie where it fell. To him it is all the same what disease the animal died of,—he only knows he is so much the poorer.

By inquiries I attempted to find out whether any particular state of the atmosphere was supposed to have a greater or lesser effect on the plague during its ravages, but met with very few men who had paid sufficient attention to the subject to be able to give any correct information. Mr. Schliept, of

Khvalinsk, says great heat increases the danger. The droughtiness of the climate is very much aggravated at times by the S. E. and S. winds, which in some seasons blow strongly for a few days, scorching the vegetation, and overpowering both man and beast. If the plague be raging at a time when these hot winds blow it appears to receive a fresh stimulus from them, and to destroy with redoubled violence. At Oosolye, on the estates of Count Orloff Davidoff, I was told that a wet autumn, such as that of 1877, would probably tend to propagate the plague in two ways : in consequence of the increase of damp situations, and their miasmatic effluvia, the cattle become more predisposed to its ravages ; in a wet autumn the peasants do not always succeed in getting in all their hay and straw—much of the former is left in the fields to be covered by the snows. They have, therefore, to go about and buy ; and this additional movement on the part of the people from one village to another aids the spread of the disease. On the whole, wet weather appears to increase the evil effects of the disease. At Alexandroffka (in the district of Boozoolook), the steward, Mr. Balk, and the veterinary surgeon, Mr. Hiegel, on Mr. Stobayoos's estates, prevented the disease entering from a neighbouring village (where almost every animal died) by strictly prohibiting all communication with that village while the plague was present, thus sending it off in another direction.

But it is not always that men act so energetically as these two gentlemen did.  In the south-eastern steppes some landed proprietors take precautionary measures, in so far as that they do not allow any cattle to travel on the roads across the estates unless one of their bailiffs has previously seen that the animals are at least to all appearances healthy.  This measure is, on the whole, but a poor one; stricken animals do not show the symptoms early enough to prevent infection, and to pick one out of a large herd, consisting of several hundred, is no such easy matter.  Inoculation of the cattle with the virus of the disease has long been given up as both impracticable and inefficacious.  At the Kazan Agricultural Training Institution, isolation, together with disinfectants and antiseptics, are the only means resorted to.  Here all the animals are isolated by twos.  If one of the two get ill, both are separately removed to another situation.  Every pair has an attendant, who is not allowed to leave them day or night, and food is brought to each of these men separately.  They found that it was only by the strictest observation of isolation that it was possible to prevent the disease spreading.  Mr. Coom, at Orenburg, allows the peasants to drive their cattle into and isolate them in the forests when the plague is about.  At first the peasants declined the offer, fearing some *arrière pensée* on his part, and they lost almost every animal.  Another time, however, they isolated their cattle in the forest, and lost only fifteen out of over a thousand head. Mr. Kœnitzer, at Samara, once saved sixty per cent. by simply and energetically attending to strict isolation : of those attacked, eighteen per cent. recovered. It is, however, very rare indeed that one is so successful. During another outbreak he killed six that were diseased at the very outset, hoping thereby to stamp it out. Owing to the wet autumn it was exceedingly difficult to keep them well watched ; but the frost came on, and the ravages of the disease suddenly stopped. Whether the frost was the real check, or whether the disease had already done its worst, is hard to say, seeing that only four per cent. of the whole herd was left. Since isolation is the only sure means of prevention, but in some parts it is not always practicable, he proposes next time to inoculate all cattle so placed, in order to have the anxiety and trouble over quickly. The steppes, when one has sufficient land at one's disposal, are very well adapted for isolation ; but during wet weather, hay-harvest, or threshing, when the animals are liable to suffer from wet or cold, or when many people are about, these steppes become useless ; and those farmers, situated in more northern parts, possessing large tracts of forest land are better off, as the cattle can then be securely watched, sheltered, and isolated.

I once heard of a very curious measure, employed by stock-keepers in the south-eastern steppes, for preventing the spread of the disease when animals are already slightly infected. They drive the cattle about as furiously as possible, harass them, frighten them, goad them across the plains and back again, driving them almost wild. When the poor beasts have perspired so thoroughly and are so dead tired that they can barely move, they are then left alone for a short time, and are said to recover altogether. For the above I will not vouch, seeing that I had it only by hearsay; but a large landed proprietor, Mr. Vadim Ossorghin, of Spakhovo (district Boosoolook), related a very similar case which happened once to him. One winter the plague broke out amongst his herd, and fresh animals were succumbing daily. One night when the cattle had been left in the forest for shelter, the guards having gone home instead of remaining to protect them, the cattle were attacked by the wolves, who drove them through and round the forest in all directions (as was plainly proved by the marks in the snow the next morning), and altogether tired and frightened the poor beasts to such an extent that at daybreak they were totally incapable of movement, and lay spread about all over the fields in a most helpless condition, with protruding tongues, and almost dead with fatigue. After that fresh attacks ceased, and soon the disease wholly disappeared.

When in 1866 the cattle-plague committed its destructive ravages amongst the prize herds in Eng-

land, many eminent men were of opinion that its
virulence was augmented by a great predisposition
to the disease caused by high-feeding, and the want
of stability arising from early maturity. With all
due deference to those gentlemen, I am not of
that opinion, as I find that information on the subject does not lead to the above conclusion. Russian
cattle belong to the most hardy animals on the face
of the earth,—high feeding is unknown to them, and
they do not arrive at maturity until the third or fourth
year ; yet when the cattle-plague does break out
amongst them, in any particular locality, it appears
to sweep away the *whole* herd.

———

## CHAPTER X.

### SHEEP.

SHEEP play an important part in the economy of
the Eastern Steppes. The large or rather vast tracts
of natural pasture, entailed by the Steppe-system of
agriculture, would in most cases bring the landowners
no returns were they not available as sheep-runs. The
market for disposing of the hay is not large enough ;
and, apart from the fact that cattle-keeping is too
risky in face of the ever-recurring cattle-plague, sheep
will generally find sufficient, even in a dry season,

where cattle would starve. Much pasture is occupied by troops of horses; but unless favoured by local influences, such as the vicinity of the German colonies, etc., they do not as a rule offer such good profits as sheep - runs, owing to the great competition with Siberian horses from over the border. Sheep consume also the greater part of the straw, which but for them would be left to rot on the fields, or burnt merely for the sake of getting rid of it.

The native breed is a very poor one, and in its pure state is only to be found on the lands of the peasantry. It is, however, very susceptible of improvement, and crossed with superior and larger breeds, has been found to become fairly remunerative. This is, however, not done so often as might be, since the largest flocks consist chiefly of Merino, or the fat-tailed Tartar (Khirgiz) sheep. The value of these latter consists chiefly in their fat; large numbers are annually driven in autumn to Samara, there to be boiled down for tallow. Their fleece, on account of its excessive coarseness and the little curl in the wool, fetches a very low price.

As to the management of these Khirgiz sheep I have no information, but I have every reason to believe that they are kept in as primitive fashion as the Tartar horses.[1] Distinctions between Elec-

---

[1] Khirghiz crossed with Merino will, after 3-4 crossings, entirely lose its characteristics, but the result will be a larger sheep.

toral, Negretti, and Rambouillet Merino are not strictly adhered to, although I was assured that most of the imported rams are pure Rambouillets. Breeders do not as a rule seek to attain any extra fine quality of wool, but rather a moderate coarse quality, and much attention is paid to obtain well and evenly-covered bodies. The quantity of wool varies from 6–9 lbs. a head. The Merino, being a comparatively small animal, is occasionally crossed with good effect with the Tcherkess, or other improved Russian breed. The management of Merinos is much as follows : when the grass appears in spring, they are driven out to pasture, remaining in the fields all the summer. At night time they are put under cover ; but where the run is very large, and consequently the distance home too great, permanent hovels or enclosures are erected in sheltered localities, which come in very usefully during inclement weather. For the winter they are kept in uncovered yards, surrounded by sheds, with easy access to water. Their food during winter consists of hay and straw. Some sheep farmers prefer keeping their sheep altogether under cover during winter, where the heat escaping from their bodies, being kept close and quiet, keeps them tolerably warm, and less fodder is required. This, however, is not considered good practice, as the sheep suffer very much from the closeness and the confinement ; they are rendered more susceptible to inflammations when driven to drink the cold water (39° F.) out of the ice-bound river or pond; and lastly, the wool is said not to grow so well as when they are kept in the uncovered yards.

Where the flocks are large, and only few rams are kept, the covering necessarily extends over a long period; and in consequence lambs are commonly dropt between February and May,—the Merinos, as is well known, seldom bearing above one lamb at a time. It is not customary to give the lambs any extra or special food when weaned, although I know of cases where oats, linseed-cake, and even hemp-cake are given.

With regard to the diseases which prevail, I could get little or no information. During hot summer days, and more especially during the predominance of the south-east wind, which makes the heat almost unbearable, many sheep are affected with and succumb to apoplexy. This disease is very prevalent all over the country; it causes the greatest havoc among the finest sheep, which are ever the first to drop. In answer to frequent and reiterated enquiries, I have not heard of any case in which cattle-plague ever attacked sheep. I was told that sometimes during its prevalence sheep become unwell, refuse to eat, and mope about, yet none died from the disease.

I occasionally heard complaints that although sheep-farming was at one time a lucrative investment, it has, in consequence of drought and the emancipation of the serfs, ceased so to be, and the country is said to be by that much impoverished. Whether it is statistically true that the country has fewer sheep than formerly I know not; nevertheless, a decrease in their number may have been brought about by the Act of Emancipation, which necessitated a large transfer of land from the landed proprietors to the peasantry, by improved communication (railways), which makes corn-growing more profitable, and lastly, by the continued stream of peasant immigration from west to east, all of which forces, by leaving less pasture available, may no doubt have also tended to decrease considerably the number of sheep. An ordinary ram fetches commonly 10 roobles; a sheep 4–5 roobles.

Goats are kept by the Tartars only. In Orenburg, and to the south-east of Samara, camels and dromedaries are used as beasts of burden by the Bashkeers and Kirghiz. Swine and poultry abound; but their flesh is tough, and they offer no points of interest. Pigs are sold by weight; 2 roobles per pood is a common price.

CHAPTER XI.

THE PEASANTRY, AND THEIR HOUSES.

THE peasantry are as a rule wretchedly housed. In erecting their houses (log-huts, cabins), the chief consideration is to secure good protection against the cold of winter, to attain which all sanitary precautions are disregarded. The cabins are built of logs, and are usually one or two roomed, and are built throughout the country essentially on the same plan. The accompanying plan gives a tolerably accurate idea of a single-roomed cabin. I took its measurement once, while waiting for post-horses at Serganka (Nickolayeff district). It contained slightly under 2000 cubic feet, and was occupied by six adults and several children. It was a fair representative of a Russian peasant's abode, and the number of inhabitants was nothing uncommon. The windows, generally three in number, are very low and small, each about 2–3 feet square ; but on account of their dinginess the amount of light they let in is insufficient. The door is also low. By low doors and windows the warm upper stratum of the atmosphere in the room is more effectually retained, and when the door is opened but little can escape. The brick stove in the corner occupies almost one tenth part of the room. In a line with the stove is the kitchen, overlooked by one of the windows, and partitioned off from the main apartment by a thin boarding. Cabins with two windows only seldom have this partition. Opposite the door in a corner is the Oratory, containing the portraits of the houschold gods, to whom every Russian bows and crosses himself several times on entering, even before he pays any attention to the persons in the room. That part A on the sketch-

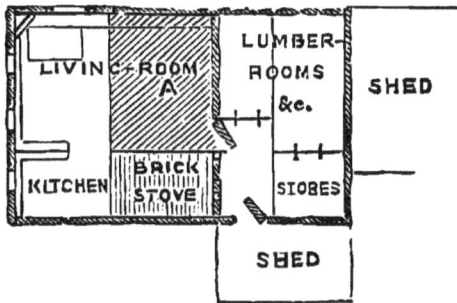

PLAN OF A HOUSE AT SERGANKA, NIKOLAYEFF DISTRICT.

Height . . . . . $9\frac{1}{2}$ feet
Breadth . . . . 15 ,,
Length . . . . . 15 ,,
——— = $2137\frac{1}{2}$ cubic feet.
Less space occupied by stove,
$7 \times 4\frac{1}{2} \times 6$ . . = 189 ,,

Atmospheric capacity . $1948\frac{1}{2}$ cubic feet.

To face page 61

plan marked off by the dotted line indicates the position of the palata. The palata consists of a few planks (elevated and fixed $5\frac{1}{2}$–7 feet from the floor, leaving 3–2 feet clear space) below the ceiling, and is used as a sleeping compartment during winter. A table in front of the Icons, a few benches round the room, a weaving and spinning machine, a suspended water-jug, some faded prints of the Crimean war, and a portrait of the Emperor, make up the furniture of the apartment. The ceiling consists of planks fixed closely together, the outer side of which is thickly covered with sand or dry earth, also straw or hay. The crevices between the logs that form the walls of the cabin are filled in with moss, and are besmeared with clay on the approach of winter. The high sloping roof, often of wood, is mostly a primitive thatch; the straw is rudely laid on about two feet thick, and firmly attached by means of transverse strips of wood. The floors of cabins are all raised a foot or so above the ground. This would expose them to the cold, which could thus easily enter from below, did the peasants not surround their cabins by manure and soil closely packed against the walls. The cabins themselves have always a damp, musty, unpleasant smell.

To decide whether the habitations are as a rule clean or not is rather hard. I have seen quite as many clean and tidy as I have dirty and untidy abodes. Every house has its own yards—an inner

and outer yard—surrounded by one long shed, running all round.   In some villages the granaries are
situated in a row down the street; but oftener every
peasant has one or more in his yards.   The yards are
enclosed by clay-plastered wattle-hedges or rough
paling, which freely permit the wind to blow in on all
sides, just as the whim takes it.

RUSSIAN MUD HUT.

Along the Volga and other districts, where good
logs for building can be obtained without difficulty,
and consequently cheaply, the log-huts are well-built;
but where the logs have to be drawn a hundred
miles and more across country, the expense is great,
and considerably increases the price,—even doubles
and trebles it.   The poorer peasants cannot afford to
buy them : they therefore build mud-huts out of air-
dried bricks.   Such mud-huts do not look very
inviting, and I was agreeably astonished to find them
on the whole comparatively comfortable.

Several families live in one and the same cabin
(generally parents and their married sons),—one

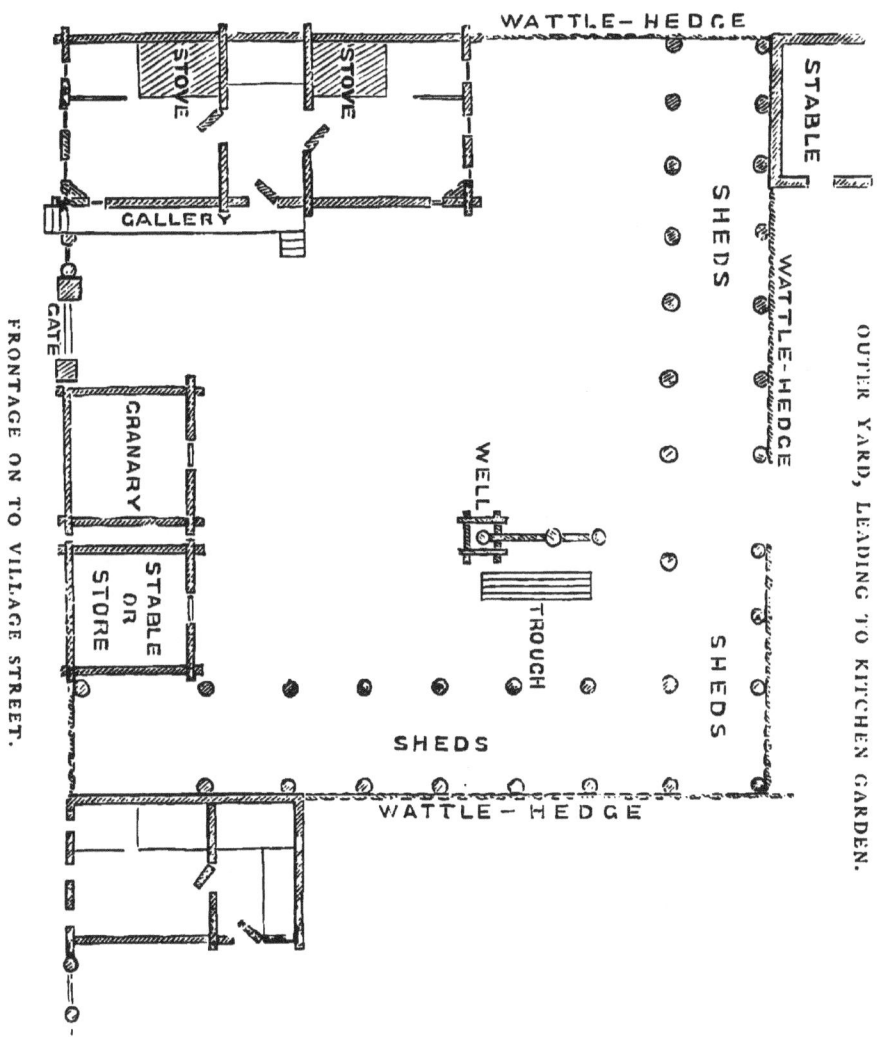

WATTLE—HEDGE

STOVE

STOVE

STABLE

GALLERY

SHEDS

WATTLE—HEDGE

GATE

OUTER YARD, LEADING TO KITCHEN GARDEN.

FRONTAGE ON TO VILLAGE STREET.

GRANARY

STABLE OR STORE

WELL

TROUGH

SHEDS

SHEDS

WATTLE—HEDGE

GROUND PLAN—PEASANTS' HOUSES AND YARDS.

To face page 63.

family occupying the palata, another the stove, and most likely a third the benches at one end of the room. The overcrowding to which the Russian peasants are so accustomed has most probably its origin in the complex co-operation which existed among the members of a peasant's family during serfdom, and which was fostered by the old proprietor in consequence of the greater influence it gave him over his serfs. In this peculiar institution all the members of the family throw their earnings into one common fund, no matter how they were earned. The members all live in one house,—the sons marry, and bring their wives to live with them in their parents' home. They can thus live more economically, are better secured against bad times, and enjoy besides, to a certain extent, the benefits of division of labour. The parents are, moreover, insured against old age, for the sons never set up house themselves until the formers' decease. These reasons have caused this style of peasant-living to be highly extolled, notwithstanding which it is slowly but surely disappearing. As a good worker is a useful addition to such a family, the father selects a good strong wife for his son (who is considered marriageable when about nineteen years old), but the young woman is to all intents and purposes more her father-in-law's slave, than the wife of the young fellow she married. Formerly, if the son rebelled, the father could appeal to the master, who of course enforced his wishes. Young men now-a-days prefer to have their wives and homes to themselves, and there are now no masters to prevent them.

The great objections against these large households lie chiefly in the indecency and gross immorality which must necessarily ensue, when several families consisting of men, women, and children of all ages live together in one apartment,—objections strong enough to counterbalance the apparent economic advantages (for that which is immoral cannot be economic), and to warrant their total abolition. Happily, as I have already mentioned, the peasant feels this himself. Living with his wife alone he has to work harder; but it is doubtless a healthy sign that he should prefer a little harder work for the sake of a little more independence and comfort.

Paradoxical as it may appear, the Russian peasant strives, as much as lies in his power, to keep himself clean. He scrupulously takes a steam-bath once a week: to enable him to do so, work is broken off rather earlier than usual on Saturday afternoons. During summer he frequently bathes twice a day. The influence of these baths, from an hygienic point of view, is great, as they are a universal institution. No village is without one; and no peasant would think, nay dare, to go to mass on Saturday night without having previously bathed. We are accustomed to sneer at the dirty Russian peasant, but I think we may safely ask, how often does one meet with an English agricultural labourer who has a bath even once in a twelvemonth?

# CHAPTER XII.

## THE RUSSIAN VILLAGE COMMUNE.

WHEN the serfs were liberated in 1861, each village commune was to receive at most five fiscal dessyatins (13½ acres) per male member, the option being left to the commune to take less if it so preferred. Unfortunately, there was at the time a notion abroad that later on the peasants were to receive the whole of the land ; and so deep a root did it take that notwithstanding all the coaxing of the arbiters [1] and ci-devant masters, the greatest difficulty was experienced to get the peasants to accept of at least one dessyatin (2·7 acres), and in a few cases even the land on which the village stood. For the land seceded to the peasants the former proprietors were, to a certain extent, only nominally compensated, the debt thereby incurred by the Government is calculated to be paid off by an annual rent of five roobles per dessyatin for fifty years.

Owing to the communal nature of the holdings, the Russian peasant-proprietor does not succeed to those full benefits that accrue from peasant-proprietorship

---

[1] Arbiters were gentlemen specially employed to aid the fulfilment of the Act of Emancipation, and at the same time to arbitrate in cases of dispute between the ci-devant masters and men.

in other countries. The village commune received the land in trust for its male members, to each of whom it gives an equal share ; but through male additions to every family, the male members consequently vary in number, and a redistribution of the land must necessarily follow, in order to equalize the size of the holdings. The consequence of this is that as the population increases the size of the land belonging for the time being to the individual peasant decreases. As is well known, every village community in Russia manages its own affairs, the most important of its functions being the collection of the imperial and provincial taxes, and the distribution of the communal land above referred to. The time allowed to elapse between the redistributions is not fixed. As a rule, I found that in localities tolerably thickly populated, where the land is rapidly approaching its lowest capabilities, and the three-field system has been introduced, redistribution takes place once in three years ; whereas, in localities thinly populated, and where no regular system of cropping prevails, redistribution takes place almost annually. When a distribution takes place, the land is divided into as many lots as there are males in the community, and to each family is assigned the number of lots according to the male members that compose it.

In meting out the land, distance from home, loss through roadway, spring-marsh, quality of soil, are all justly considered. An equal division of the land appears to entail a great drawback in obliging members to accept of small irregular plots, situated in various parts of the communal land, often at a great distance one from the other. The frequent distributions of the communal lands do not materially hinder the execution of agricultural operations, although occasionally I found that spring-sowings had been retarded on account of the partition not having been carried out in time.

A former and strong objection to this system of land-tenure was that it precluded progress, inasmuch that the peasant, knowing his piece of land will most likely change hands at the next distribution, refrains from improving it. To manure land is one step in the path of progress; and in order to ensure to a man the benefits thereof, a law was enacted four to five years ago restricting the distribution of the land to every twelfth year in those localities where manure has come into use.[1]

The ground occupied by the village and roadways, together with the permanent pasture or fallow, leaves so much the less arable land : thus each peasant rarely cultivates more than two-thirds of the amount allotted to him. A few villages to the north of the province of Samara possess forest land. Many villages set apart a piece of land for pasture ;

---

[1] At Doobovoy, where *no* manure is likely to be used for the next ten years, the Elders of the Commune divided their land last year, and decided not to do so for another twelve years because it is such trouble.

but most rely upon the neighbouring landlords for their fodder supplies. The commune places no restrictions on the peasants as regards the cropping. Where the population is thick, the three-field system is pursued; where the population is scanty, the land is cropped unsystematically. Poor peasants not possessed of that small capital necessary for the exploitation of their share, or such as prefer town-life, sublet their land to others, who are usually glad to get it.

In the Boogoolma, Boozoolook, Nikolayeff, and Novo-Oozensk districts there are still large tracts of State lands which are farmed by the peasant (late State serf), in the same way as though they belonged to the Commune. I thus found that a peasant sometimes cultivated shares of above twenty dessyatins (fifty-four acres). As the land gets taken up these large shares are gradually reduced, and will be ultimately reduced to the usual quantity of five dessyatins.

Without entering into a discussion on the relative merits or defects of Russian peasant-proprietorship, I wish to notice one point concerning its future influence on the manufactures, and which point, I believe, has not hitherto received any attention. The Russian Government strives, by a prohibitive tariff, to foster its home manufactures. When manufactures are so far established throughout the country that a brisk competition has set in among them, the manufacturers will be obliged to lower the wages ; but if these lowered wages should amount to less than a peasant can earn comfortably by cultivating the share of land he is entitled to in his native village, will he not *en masse* return to an agricultural life ?

## CHAPTER XIII.

### THE LABOURER.

THE difficulties experienced with the Russian peasant and labourer are still very great. The chief charges brought against him are indolence, drunkenness, and dishonesty. On the whole, however, he is not such a bad fellow as report makes out : to bondage, from which he is just emerged, may be ascribed the greater part of his vices, although climate and other influences have borne a large share in reducing him to his present wretched condition. During the whole winter there is an almost absolute want of work. A peasant may find a little wood-cutting to do ; now and again he may be commissioned to transport grain to town, or occasionally he threshes out his rye ; but the greater part of the winter is dozed away on the brick-stove at home, where he consumes the earnings of the past summer. Generally, the land he scratches up is sufficient to support him and his family, so that there is little to stimulate him to greater exertion. Transporting grain from the villages to town is to the peasant a comparatively light and remunerative occupation, and whenever he can obtain such work he refuses all other ; it is even not rare that when offered high wages, with hard work, he will, after a short trial, give it up, preferring his pittance and his ease.

I have often watched the peasant at work for himself; and although he appeared to be at great pains for the attainment of his object, he always did his work in such a drowsy sort of way that the small result appeared to me out of proportion with the amount of labour bestowed. Last year the crops throughout the country were very good. One would think that after the many failures the peasant has experienced, he would have been most eager to gather them in; but no, he procrastinated until heavy rains set in and destroyed above one half his crops.

A man who does not work well for himself cannot be expected to do better for others. Hired labour is, therefore, at a very low standard. In most cases, however, I should say it depends upon the bailiff or employer whether the peasant works well or not ; if well treated, and liberally supplied with good food[1]

---

[1] Of late it has become customary to give meat at the rate of half a pound per day, and it is said with a good result to both master and man.

(a point concerning which he is most sensitive), I have known him to get through a remarkable amount of labour, and that, too, willingly and with pleasure. By treating the peasant well, I mean kindly but firmly, and above all, justly. Bailiffs in Russia are proverbially a bad lot, and their knavery is the primary cause of the greater part of the trouble that arises between master and servant; they are most overbearing and unjust in their conduct, and are fond of large doses of abuse; whereas few people are more susceptible to a kind word and fair dealing than the Russian peasant. Frequently, when pressed to complete work by a stated time, I have succeeded in my object by simply appealing to the good-will of my men, whom I always found ready to respond to such an appeal. Certainly a reproof in season is useful in steadying the natural instability of the Russian character; but it may be administered without any of that scurrilous invective the Russians are so fond of.

The drawback caused by drunkenness to industry is very great, notwithstanding which the peasant does not drink so much as is generally imputed to him. He certainly does not drink so much as the average Englishman. We hear a great deal about his drunkenness, because, owing to his backward and ignorant state, he possesses little self-respect. An Englishman when drunk is ashamed of himself, and sneaks away home; not so the Russian peasant, who proclaims it joyously to all the world. I have seen him occasionally come to his work of a morning the worse for liquor; but he will set resolutely to work, and boast that, notwithstanding his inebriated state, he is able to do his duty.

In countries like Germany, where beer is consumed in large quantities, much less is heard of the evil effects of drunkenness than in countries like England and America, where ardent spirits form the principal drinks. The same of Russia : vodka (rye-whisky) is the only spirituous beverage known to and attainable by the peasant. Not being able to buy much on account of his poverty, he gets the easier drunk when he does drink ; and, as is well known, the effects are worse than were he to drink beer. Old orthodox Russians look upon vodka as an instrument of the devil, and refuse to touch it; others will only drink when given to them, and many will not spend their money gained by ordinary labour on vodka, although any exceptional windfall is generally consumed that way.

The holidays, of which there are unhappily only too many, are as prejudicial to steady application as they are to the morals of the peasantry, who, having absolutely nothing else to do on such days, must needs resort to the dram-shop. Some villages are notorious for their drinking propensities. I know of one village (Bolshoy Glooshitza), containing over three thousand inhabitants, where much vodka is consumed on account of a labour-fair which takes place once a week. The money obtained for spirit licenses covers all the taxes the village has to pay, and leaves a small balance, which is given to the widows and the fatherless. I am told this happens also in a few other instances.

The Russian peasant is a very great liar; dishonesty and dissimulation form part of his trade ; he feels not the slightest compunction in telling the most absurd and flagrant lies when called to account for misbehaviour. His notions of the mine and thine are singularly elastic so far as agricultural objects are concerned. He is very quick at finding pieces of iron, etc., before they are lost. The carts and greater part of the implements of the proprietor are almost the same in size and pattern as those of the peasant, and are, therefore, not easily identified : in consequence of which the latter personage is very apt to mistake those appertaining to his master for his own property. The chief difficulty with him is his cunning in stealing hay, allowing his cattle to trespass, and pilfering wood from the forest, in all of which I have had far more experience with him than is absolutely pleasant. His excuse for stealing hay and wood is most naïve : God, he maintains, sows the grass and plants the trees, therefore he has as much right to them as any man !

Thefts of the above sort, when discovered and punished, will now and again lead to incendiarism which unhappily is still tolerably prevalent. The frequency of fires from various accidents, and the inflammable nature of the material (wood) of which all buildings are constructed, causes quite sufficient loss without the additional destruction of property out of pure brutal revenge. The investigation of a case of incendiarism is impossible. When a peasant gets into trouble he swears he was drunk at the time, and therefore could know nothing about it ; and to attempt to arrive at the truth through a bailiff is simply to give the man a fresh opportunity to perjure himself. Besides the fine for stealing hay or wood, there are many other fertile sources of incendiarism. Very often short wages are paid, or a bailiff receives a bribe from the peasants, and then does not act up to it ; or a bailiff and a peasant, having combined to rob their common master, fall out : the former, being the more powerful, comes off best, and the latter out of revenge sets fire to his master's buildings. But whether the peasant is right or wrong in his grievances incendiarism is still too rife an evil ; the wonder is the greater that such acts still occur, for the remedies lie close at hand. It is not sufficient for a proprietor to live on his land, but he must also take a more lively interest in its management, which will counteract the dishonesty of the bailiff ; this is the best preventative to discontent.

When any buildings have once been burnt down, the landholder can raise the rent of any pasture he lets to the neighbouring commune to such an extent as will compensate him for the loss he has sustained. This will deter the recurrence of the act by making it in fact a crime against the welfare of the commune, which in the eyes of a Russian peasant is a sacred institution. This latter remedy was first tried by a gentleman who has the management of some large estates in Orenburg,—he found it to act infallibly. My sojourn in the country was too short to enable me to satisfy myself personally that the peasants had made any progress of late years ; many gentlemen told me that the progress was small, and not near so great as had been expected. However, the Russian peasant is in such a low, unenviable condition that any progress, no matter how small, is of serious importance. His servility, grovelling, his gross ignorance and superstition form, in fact, the most melancholy feature throughout Russia. The peasant carries the seed of improvement with him : he is exceedingly inquisitive, is always craving for information about foreign countries, and an explanation of some new machine affords him immense satisfaction. When travelling I have occasionally been surrounded on an evening by a crowd of peasants who paid the most profound attention to my descriptions, or indeed anything that to them was new, and who showed their interest by asking for more and more details.

Among those who speak well of the progress of the people are captains and engineers on the Volga and Caspian steamers, who being foreigners are less likely to be biassed in their opinions. They bear witness to a man that the Russian peasant is already a different being to what he used to be, that he is gradually but surely emerging from his old lethargic state, and that he will work more readily than formerly when the knoot forced him.  One captain in particular remarked that whereas sixteen years ago few could read, now-a-days it is a common sight to see several peasants sit down together on deck with a book which each man that can, reads in turn.  There are many Germans from the Baltic Provinces, who were estate managers before bondage was abolished, whose testimony is to the same purport, namely, that the peasant works harder, is more careful, and owing to his freedom takes a greater interest in his own affairs.  The more civilised costume at present worn by the women ;  the more general use of the national tea-urns (samovars), of which every village now possesses several, whereas formerly it was difficult to find one ;  the use of Kerosin lamps, which take the place of the old prepared Verbascum and splint lights, and many other little particulars indicate an improved material welfare. But sad to say, morally everything is still left to be desired.  The Russian clergy do absolutely nothing to aid the people either materially or morally ;  they are too greatly imbued with the importance of their personal prosperity to trouble themselves with the people, and in many villages they go so far as to set a highly vicious example themselves.

There are many Russians who deny any progress on the part of the peasant, because, as they maintain, he drinks more than formerly. Even were he to drink more, it would prove his progress more conclusively, for to drink more his affairs must be in a better condition now.

There are, strange to say, still peasants who wish the old days of bondage back again : they complain that now when they want a thing they have to buy it ; when a horse dies, or hay, corn, or fuel is required, they have to work for it, whereas formerly the master gave them all necessaries. Such men must have had very lenient masters.

The Government, chiefly the Provincial Assembly, has had a good share in the attempt to ameliorate the condition of the peasant. The establishment of primary village schools has already been alluded to. A good system of medical assistance has been introduced, coupled with the foundation of hospitals. There is further a special insurance-tax against loss by fire, which the peasant has to pay. In a country where fires are so common the wisdom of compulsory insurance is undoubted.

So far I have not mentioned the Tartars. They are not often regular labourers in the province of Samara ; Kazan and Orenberg are their homes, but they come in for a large share of hay and harvest work throughout Samara. They are excellent labourers, honest and sober, and having never been bondsmen, are untainted with the vices the Russian peasant contracted under serfdom.

G

## CHAPTER XIV.

### RURAL ECONOMY, WAGES, WORK, AND TAXES.

THERE are several old families who possess very large estates, containing above 50,000 fiscal dess (81,000 acres), though generally speaking an estate of 5,000 dess. (13,500 acres) is considered a large one. In the absence of any law of primogeniture it is customary for the land, on the decease of a proprietor, to be divided equally amongst his children, a procedure which, in a few generations, minces the largest estates.

Fifteen years have now elapsed since the Act of Emancipation was promulgated, and complaints are still loud and numerous about the hardships it has brought upon the landed proprietors. A change which in fact took away their working capital was a change of too radical a character not to have affected the landholders most intimately, but it was not the sole cause of their ruin. The demoralising influence of serfdom seems to have taken all energy and self-denial out of them ; thus, when capital was obtainable from the Banks, in the form of mortgages, the greater number knew not how to apply it, and then reckless expenditure often helped to accelerate their downfall. From this some now maintain that the banks shared equally with the emancipation of the serfs the ruination of the landholders.

Estates situated far from the Volga or Kama suffered most, as not only does the cost of transport increase directly with the distance, but at the same time the cost of production as well,—owing to the scantier population the further the remove from the above means of internal communication.

At Orenburg, for instance, cultivation no longer paid, and the land was either let to the peasant for cultivation, or to the Cossacks, Bashkeers, and Tartars, for pastoral purposes. Since the opening of the Orenburg-Samara Railway, eighteen months ago, this has considerably changed, and the cultivation of the to all intents and purposes virgin soil has made agriculture once more a remunerative employment.

As a rule, the landed proprietors cultivate their own land, either personally or through their agents and bailiffs. Where this is not done they let the land in driblets to the peasants, who generally rent the fallow as pasture, paying per acreage or per head of cattle pastured. There is, therefore, no such class of tenant-farmers as we have in England ; but occasionally richer Germans from the colonies along the Volga at Saratoff rent large farms, either on long leases or yearly tenancy. There is generally a stipulation that the land will be allowed a certain amount of rest at stated intervals.

Unfavourable seasons, like those of the years 1872,

1873, 1874, and 1876, have brought agriculture into bad repute, and induced many old proprietors to sell their land. To many of them the loss caused by a bad year is irreparable, as they are unable to curb their expensive habits, and borrow to make up for loss ; while, being always short of ready money, they are frequently obliged to send their produce to a glutted harvest-market, at times selling at a loss, and thereby foregoing the advantage of waiting and withholding the grain until the market suits them.

In both these respects the peasant also suffers severely. He is generally obliged to sell his produce directly it comes in ; and worse still, after a bad season, he has not sufficient grain to sow the next crop. This I noticed particularly last year. The harvest of 1876 was in various districts a very bad one, and in spring, 1877, many a peasant fell short of seed : they therefore sowed whatever they could scrape together—oats, wheat and rye mixed, or very dirty seed. But a crop grown from such seed, although very good for home consumption, will not pay when sent to market.

Land put up for sale is chiefly bought up by the merchants, who in conjunction with their capital have steady business habits, which enable them to overcome many an obstacle insurmountable by the ordinary landed proprietor. As a proof of the business capacities of the Russian merchant, may be cited the general absence of the Jews, who from Moscow eastward are unable to make any headway against them. The superior general order and economy of arrangement of a merchant's estate is quite striking. He is, moreover, too shrewd a man not to be aware of the importance of keeping books, an arrangement otherwise unknown to agriculturists.

A Russian merchant seldom or never takes the initiative; for instance, if a new machine comes under his observation he will wait and convince himself that others can make it pay before he adopts it for himself,—he will not buy a new machine on the presumption that its introduction is remunerative. This is in deep contrast to the old proprietors, who buy either recklessly or are profoundly indifferent.

In the Novo-Oozensk district, merchants and rich peasants do a large business in speculative wheat-raising. They rent a few thousand acres of fresh land for a twelvemonth; about June or July they send men down in batches from Samara, Simbirsk, and Kazan to plough it; and in spring they again send labourers down to sow and harrow. At harvest-time occasionally batches of hundreds find their way to do the work : the crops are harvested, the grain threshed out by machine or horse, and sent to the granaries in Samara, all within the space of a few weeks. Much money is made in this way, although the hot dry winds often destroy many a fine crop. Success depends chiefly upon the quickness with which the grain is harvested, threshed and brought to town; for rain may come down, hinder work, and the expense of maintaining and keeping together the men out of work is so great as to swallow all profit, and cause loss as well.

When a fair harvest is expected, labourers from Simbrisk, Kazan, and even Vyatka flock to these districts, occasionally in such numbers that the competition thereby engendered reduces the rate of wages to a minimum. This was the case in several localities last year. The land for the greater part belongs to the Government, who let it at a nominal rate to a few peasants, who in turn sublet it with large profit.

It is not at all extraordinary for a man to have from ten to fourteen thousand acres under White Turkish wheat—four thousand acres is a very common amount; and I know positively of one rich peasant who cultivated in this style last year over 28,000 acres. An English farmer can imagine what it is to see such vast tracts waving with golden grain.

Many landed proprietors, notwithstanding the possession of hundreds of acres, keep no permanent labourers, but a bailiff and a groom. They keep no horses but for personal use, no bullocks; and have neither ploughs nor harrows,—in fact, no implements whatsoever, and no granaries,—all the agricultural operations being performed by the labourers hired with their own bullocks and implements. They maintain this to be the only paying system. It is very common. The following is a statement of cost of cultivation under above circumstances :[1]

| | | |
|---|---|---|
| Ploughing ... ... | 6·oo roobles per sorokaya dessyatin | |
| Sowing and harrowing | 3·oo ,, | ,, |
| Seed ... ... ... | 6·oo ,, | ,, |
| Threshing ... ... | 6·40 ,, | ,, |
| Cutting and binding ... | 8·oo ,, | ,, |
| Rent calculated at ... | 6·oo ,, | ,, |

$$35·40$$

Crop...80 poods at 60 R. gives 48 R.

Less .. ... ... 35·40

Gives a nett return of ... 12·60 roobles per sorokaya dessiatin, 9s. per acre ; but 10 R. nett return or 7s. 2d. per acre is considered the mean. This is in western Boogorooslan district.

The following is an extreme estimate for cultivation of wheat in a very favourable year, in Novo-Oozensk, under like above circumstances :

| | | | |
|---|---|---|---|
| Ploughing ... ... | 6 roobles per sorokaya dessiatin | | |
| Harrow ... ... | 2 ,, | ,, | |
| Cutting and Binding | 25 ,, | ,, | |
| Threshing ... ... | 16 ,, | { highest price at 20 K. per pood. | |
| Transport ... ... | 20 ,, | { great distance at 2½ K. the sack of 10 poods (=360 lbs.). | |

70 R.

The crop of which 8 sacks at 10 poods at 90 kopecks gave 72 roobles per dess., or a nett return of 2 roobles without rent of land, which was a matter of

---

[1] See also Appendix.

a few kopecks only. This is justly considered a poor profit. Very favourable years are not liked, for the market is glutted with grain, which sells low, and labour for threshing and reaping rises enormously. This is usually more especially the case with rye, as the peasants grow large quantities of this crop ; and, if a proprietor has to pay 14 roobles (or double the usual cost) for reaping alone, his year's budget will show a heavy loss.

Land is let from 1 rooble to 10 roobles per shest-daysyatty dess. (8d. to 6s. 8d. per acre). The price of land varies from 10 roobles to 60 roobles and more (6s. 8d. to 37s. 6d. per acre). In both cases the value is determined more particularly by the position as regards distance from market and thickness of neighbouring population than upon the comparative quality of the soil ; previous cropping, and the cropping to follow, also influences the rentable value.

In Boozoolook and Nikolayeff districts are some Bashkeer settlements. The Bashkeers do not cultivate their land, but wander over it nomad fashion. What they do not require for pasture they let to the peasantry. With their primitive habits they can let their land much cheaper than the Russian proprietors. The Russian peasants are much fonder of cultivation on their own account, notwithstanding all the attendant risks, than working for wages ; and thus the Bashkeers, while lowering the rentable value of land, at the same time enhance the value of labour. The Orenburg and Ooral Cossacks and the Kirghiz affect their respective neighbourhoods in the same way.

Labourers are generally hired for the year, the agreement being made some time before Easter. Although agreements are entered into, men do not think them at all binding. As summer advances and wages begin to rise, they gradually ask leave to go home on some important mission or other; and if permission is withheld, they take French leave. The prosecution of such men does not pay, as they never possess anything. Agreements with a gang of men are now generally entered into for the performance of a certain piece of work. To contract with one man for the performance of a certain piece of work, by a number of men to be employed by him, is bad,—he either cheats his men by not paying them, or they cheat him by receiving money in advance, and then sloping.

This introduction of piece-work gives better results and greater satisfaction to both parties. Time-work gives to men naturally indolent too many temptations to be dishonest, and affords them opportunities for various mean expedients to get through their time, which are not so open to them in piece-work. I have known them to overdrive their oxen in hot weather so as to make ploughing impossible; the blame is attached to the hot weather, and the distance from the homestead being great, they have in consequence to be idle. To lose or forget any indispensable part of an implement, as well as many other rude shifts, are favourite methods of getting rid of their work. To withdraw their pay is but a poor consolation, for the work remains undone ; whereas, in agriculture, and more particularly so in Russia, it is of paramount importance to have the work well forward.

Labour is as a rule cheapest in the Boogoolma and Stavropol districts. In the Samara, Boogoorooslan, and Boozoolook districts it is dearer ; and in the Nikolayeff and Novo - Oozensk districts it is dearest. Board and lodging are always included when labourers are hired for any length of time. Lodging consists in being allowed to sleep wherever they find a convenient place ; the board consists of 5 lbs. of rye bread per day ; they receive kasha twice a day (boiled millet with ½ lb. of fat, or sunflower-oil on fast days), and of late ½ lb. of meat per day has been given,—salt, and occasionally fish when no meat is to be had, are also given. The above food, together with wages, costs the employer 8–9 roobles (20s. to 22s. 6d.) per month. In money wages the labourer receives 10–20 roubles (25s. to 50s.) for the whole winter; 4½ roobles (11s. 3d.) is a common month's wages. But during harvest wages rise to 2 roobles (5s.) per day ; and if by the piece, the labourers get from 6 roobles, at Stavropol, to 25 roobles at Novo-Oozensk per sot. dess. (3s. 3d. to

14s. per acre). Mowers receive from 1 to 3 roobles per sot. dess. (6d. to 1s. 6d. per acre). Reapers and mowers receive food as well as cash payments : thus a mower receives ⅓ pood (18 lbs.) of rye-bread, ½ pood of millet or buckwheat flour, and 12 lbs. (10¾ lbs. English) per sot. dess. (4½ acres). For threshing by means of horses the peasants are paid 4-8 kopecks per pood (2d. to 4d. per 60 lbs.), employing their own horses. For steam-threshing the employer hires a company of men, who undertake the threshing collectively at about the same price per pood for wheat, and at about 2 kopecks per pood more for linseed, the employer providing stoker and driver, and machines : this price includes also the collecting of the sheaves into a stack before, or bringing them up during threshing. Night threshing is not at all uncommon. The men engaged divide themselves into day and night gangs, and go to work alternately. Such is the importance to get all threshing done rapidly, that Sundays and holidays are completely forgotten.

Ploughing costs 3-5 roobles per imp. dess. (2s. 8d. to 8s. per acre) ; but when the labourer employs his own oxen and ploughs, the cost is close upon 2 roobles per imp. dess. more. When the sokha is used the cost for labour is less; and where old pasture land is ploughed up, it is considerably higher.

On account of the lesser risks incurred, landlords prefer paying for labour in land to money. Modifications of the Metayer and Conacre system are

thus pretty widely disseminated. When labour is
paid for in wages and the harvest is bad, the landed
proprietor loses the interest and profit of his capital
invested in the land, and the money paid away in
wages as well ; but, if the peasants give their labour
for an uncertain stipulated future participation of
profit, the landlord in bad seasons loses the interest
and profit on his capital only. In Metayer the proprietor occasionally supplies the seed and rarely the
oxen, though under certain unfavourable conditions
he will do so, and even supply the peasants with a
little food as well. In the division of the profits
the peasants get from a third to a half after the
deduction for seed, etc. In Conacre generally for
every one dessyatin cultivated from beginning to end,
every labourer receives one to cultivate for himself ;
the land, however, is given to a body of men who are
made collectively responsible, with the stipulation
that the operations on the proprietor's land are in
every case to be concluded by the members of the gang
before they touch any part of the land lent in return.

As I mentioned previously, the peasantry rely principally on the neighbouring landholders for pasture and
meadow ; this places them, to a certain extent, under
the control of the latter, to whom it is a rather
advantageous circumstance. When pasture is wanted,
the proprietor exacts not money but labour in return.
He thereby overcomes any difficulty in catering for
labourers, but is also sure that his land will be cul-

tivated and crops collected at the least possible risk ; for it is always stipulated that, weather permitting. the proprietor's land is to be attended to first. When peasants are in very straitened circumstances during winter, which is not uncommon, the neighbouring landlord makes money advances[1] : he gives say three roobles (7s. 6d.) to each man in want, who, in return for that sum, has to reap a shest. dess. ($3\frac{1}{2}$ acres) next harvest. It happens that the harvest is a good one, with wages at 7 roobles per shest. dess. (5s. an acre), then the peasant is a great loser. The landlord is always more or less insured against loss, his only risk being illness or death of the peasant, for harvest wages seldom, if ever, fall below 3 roobles per shest. dess. Nevertheless, aware of the disadvantage, he suffers under, the peasant executes the work in a most wretched fashion, being only too glad to get through it anyhow, and hurrying to be in time for reaping more profitably elsewhere.

The assessment in 1876 for the Imperial Revenue was 1·1 kopeck per imp. dess. (10s. 1d. per 1000 acres). The Samara Provincial Taxes were 1·58 kopecks in the rooble on the income ($3\frac{3}{4}$ in the pound) ; and the contributions for the noblesse are levied at the rate of one kopeck per imp. dess. (9s. 2d. per 1000 acres). Besides the above, there are the "Ooyezdny," district or local rates, which are very high. A gentleman in the Boozoolook district, to whom I am also obliged

---

[1] See also Apppendix.

for much other interesting information, was kind enough to allow me to take the following extract from his papers, showing the amount he had to pay in taxes and rates. He owned 3304 imp. dess. (9020 acres), and taxes were levied on an hypothetical income of 2721 roobles.

|  |  | R. | K. |
|---|---|---|---|
| 1. | Imperial taxes at 1·1 K. per dess. on 3304 Imp. dess. ... ... ... ... = | 36 | ·34 |
| 2. | Provincial taxes and rates, 1·58 K. on an income of 2721 roobles ... ... ... = | 42 | ·99 |
| 3. | Ooyezdny, or district taxes and rates, 8·6 K. on the above income ... ... = | 234 | ·00 |
| 4. | Contributions to the noblesse, 1 K. per dess. on 3304 dess. ... ... ... = | 33 | ·34 |

R. 346·67

|  | £ | s. | d. |
|---|---|---|---|
| 1. 10s. 1d. per 1000 acres ... ... ... | 4 | 10 | 11 |
| 2. 3⅖d. on an income of £340 ... ... | 5 | 7 | 8 |
| 3. 17⅓d. „ „ ... ... | 29 | 5 | 3 |
| 4. 9s. 2d. per 1000 acres ... ... | 4 | 2 | 8 |

£43 6 6

The Provincial and District taxes include school-rates, bridge repairs,—Government contributions to the maintenance of posting-stations, etc.

# CHAPTER XV.

## AGRICULTURAL INSTITUTIONS.

DURING serfdom many estates were managed by the proprietors, aided by a more intelligent serf acting as bailiff, who had in all probability been educated for the work by the owner. The larger estates, whose owners lived chiefly in the towns, were principally managed by German stewards from the Baltic Provinces, assisted by serf-bailiffs.

When the Emancipation Law was promulgated, the greater part of these bailiffs wandered away, either returning to their Commune, or setting up for themselves, often leaving their previous owners in a very awkward predicament for the want of some competent person to manage their affairs. This want, coupled with the desire to improve the agriculture, was the cause of the establishment of Agricultural Training Institutions in various parts of the country. There are in all ten Agricultural Training Institutions in Russia: seven under the jurisdiction of the Ministry of the State Demesnes; one under that of the Minister of Education; two under that of the Provincial Assemblies. Of the seven Institutions under the surveillance of the Minister of State Demesnes, two are colleges, viz., the Petroffsky

Agricultural College, near Moscow, and the Agricultural College (translated from Poland) in Petersburg ; the other five Middle-Class Agricultural Training Institutions are : the Mariensky Agricultural Training Institution, in the Province of Saratoff; the Kazan Agricultural Training Institution, in the Province of Kazan; the Gorkakh Agricultural Training Institution, in the Province of Maghileff; the Oomansky Agricultural Training Institution in the Province of Kieff ; and the Kharkoff Agricultural Training Institution, in the Province of Kharkoff. [1]

A description of the Kazan Agricultural Training Institution, which I visited, may not be out of place here, as showing what the Russian Government is doing on behalf of agricultural education.

The grounds of this institution cover several acres, and are excellently laid out. The houses of the farm-manager, head-master, and masters, are all semi-detached wooden châlets, surrounded by lawns, shrubberies, and flower-beds, intersected by pleasant gravel-walks under the shade of the trees that so agreeably welcome the stranger from the Steppes,

---

[1] The pupils are drawn from all grades of society, as the following table shows ; they are statistical of four Agricultural Institutions only, namely, the Gorkakh, Kharkoff, Kazan, and Mariensky :

| | |
|---|---|
| Sons of the nobility, which includes those of military and naval officers, and landed proprietors... ... ... | 140 |
| Sons of the clergy ... ... ... ... ... ... | 26 |
| Sons of peasants, butchers, soldiers, and colonists ... | 189 |
| Sons of foreigners ... ... ... ... ... ... | 9 |

who, accustomed to the piercing wind or scorching sun, is amazed to find so rich a vegetation on so poor a soil, and in such a high latitude. All that is possible has been done to transform an arid sandy highland into a lively and cheerful abode. In the centre of the grounds stands the Corpus ; close by is the practice-plot of the scholars ; on the right, the kitchen garden, strawberry-beds, and orchard, the implement sheds, and repairing shops ; to the left, the houses of the masters, the farm and its buildings ; and at the back, the apiaries in the woods leading to the sunny banks of Lake Kaban.

There are at present about eighty scholars, that is to say, about twenty more than the Corpus was originally intended to receive, — the enlargement of the building is in consequence being proceeded with as fast as possible. The Corpus is a large rectangular structure consisting of ground, first, and upper floors. The ground-floor contains the domestic offices, dining-room, and chemical laboratory; on the first-floor are the class-rooms and the study, with the library ; a spacious dormitory occupies nearly the whole of the upper floor, besides a smaller bed-room for the elder scholars. There is also a lecture-room fitted-up as a museum.

Though the habits of the country are not in any way too cleanly, it would nevertheless astonish a foreigner to see the scrupulous order and cleanliness kept throughout the building. A severe cleanliness belongs to part of the educational system here. On many a large estate the bailiff lives with his wife and children, so huddled up in his two or three-roomed cabin that to keep the air pure, or the room neat and free from dirt, becomes almost impossible.

The knowledge required of a boy on entrance is slight. Entering the lowest, or first class, he is expected to know how to read fluently, and must be able to relate afterwards tolerably accurately what he has read. He must be able to write easily, and orthographically correct, and must know the four first rules of arithmetic. Boys who have passed the course of a district-school can enter direct into the second class.[1] The curriculum is as follows :

| | | | |
|---|---|---|---|
| *1st Class :* | Religious Instruction... | ... 3 | hours weekly. |
| | Russian Language ... | ... 6 | ,, |
| | Geography ... ... | ... 1½ | ,, |
| | Writing... ... ... | ... 6 | ,, |
| | Drawing ... ... | ... 3 | ,, |
| | Arithmetic ... ... | ... 7½ | ,, |
| *2nd Class :* | Religious Instruction | ... 3 | ,, |
| | Russian Language ... | ... 6 | ,, |
| | ,, History ... | ... 1½ | ,, |
| | Geography ... ... | ... 1½ | ,, |
| | Writing ... ... | ... 3 | ,, |
| | Drawing ... ... | ... 1½ | ,, |
| | Arithmetic and Algebra ... | 4½ | ,, |
| | Geometry ... ... | ... 3 | ,, |
| | Natural History (Bot. & Zool.) | 3 | ,, |

---

[1] No boy to be under thirteen years of age on entrance.

*3rd Class:* Russian Language ... ... 4½ hours weekly.
      „ History ... ... 1½ „
      Geography ... ... ... 1½ „
      Algebra ... ... ... 1½ „
      Geometry ... ... ... 1½ „
      Natural History (Bot. & Zool.) 3 „
      „ (Min.) ... 1½ „
      Natural Philosophy ... ... 4½ „
      Chemistry ... ... ... 4½ „
      Mensuration ... ... ... 1½ „
      Plan Drawing ... ... 1½ „

*4th Class:* Russian Language ... ... 1½ „
      Algebra ... ... ... 1½ „
      Natural History (Bot. & Zool.) 3 „
      Meteorology ... ... ... 1½ „
      Natural Philosophy ... ... 1½ „
      Chemistry ... ... ... 1½ „
      Mensuration ... ... ... 1½ „
      Plan Drawing ... ... ... 1½ „
      Building ... ... ... 1½ „
      Book-keeping ... ... ... 1½ „
      Kitchen Gardening ... ... 3 „
      Farm Crops ... ... ... 3 „
      Agricultural Implements ... 1½ „
      Animals of the Farm ... 3 „

*5th Class:* Building ... ... ... 3 „
      Architectural Drawing ... 3 „
      Book-keeping ... ... ... 3 „
      Farm Crops ... ... ... 3 „
      Agricultural Implements ... 1½ „
      Animals of the Farm ... ... 1½ „
      Forrestry ... ... ... 3 „
      Rural Economy ... ... 4½ „
      „ Technology (Industrial
         Chemistry).. ... 1½ „
      Agricultural Law ... ... 3 „

Not feeling myself competent to give an opinion as
to the value of the subjects taught as above enumerated, I leave it to others to decide, but should at
the same time like to add that I consider the instruction in farm machinery and implements decidedly
insufficient. My short experience in the country
inclines me to think that if many of the bailiffs under-
stood how to superintend the machines and implements of the farm, they would not so often be laid
aside as impracticable or useless, and their extension
would be considerably increased, to the benefit of the
agriculture of the country. Formerly the Institution
possessed a portable engine and threshing-machine,
but finding it did not pay, the authorities sold it.
This is a great pity, as most of the landed proprietors
possess such machines ; and a pupil, when he goes out
as bailiff, will find himself at the mercy of the first un-
scrupulous mechanic he has to deal with.

For the better illustration of the Lectures, the
Institution possesses a small but good agricultural
museum, containing models of implements, mechanical
apparati, surveying instruments, appliances for teach-
ing natural philosophy, wax models of the best breeds
of cattle, sheep, etc., anatomical figures, collections of
seeds, an excellent herbarium, samples of soils and
manures, and numerous diagrams. The library is at
the disposal of the elder boys, and contains, besides
Russian works and translations, many foreign standard
books bearing on agriculture.

Theoretical instruction is confined to the winter session (from the middle of September to the end of April), when the classes are held from 8-1 in the morning, consisting of three lessons of one and a half hours each, with short intervals at the end of each lesson. In the afternoon the pupils are free to do as they please, preparing their next day's work in the evening. Notwithstanding there is no practical instruction during the winter, yet various pupils are from time to time told off to attend to the stock, threshing, or any other work that may be on hand.

In summer the pupils receive work to do on special allotted plots, where they are taught the use of the implements, and are initiated into every species of field work, where various experiments are carried on, and where all agricultural plants that will thrive in this part of the country are grown, so that they may get accustomed to them. They are further taught the routine on the farm, attendance on cattle, and are taken on botanical and farm excursions in the neighbourhood. Practical instruction in forestry, butter-making, and apiarism[1], form part of the summer programme of work.

Forty of the pupils are maintained wholly at the expense of the Government; the others pay 125 roobles (about £15 12s. 6d.) per annum, which sum

---

[1] In consequence of the quantity of wax candles burnt before the Icons in the Greek Church, the demand for wax is very great, and apiarism becomes a very remunerative business.

includes all—board, lodging, clothing, education, and books. The fare is very good: morning, bread and milk, or on fast days honey, bread and water ; at one o'clock, soup with boiled meat, or shtchee (cabbage-soup), roast (sometimes fish), with greens and potatoes; at four or five o'clock, milk and bread ; supper, same as at dinner. In summer they receive plenty of fruit, and in consequence of the work they eat much more. Spirituous liquors are totally prohibited. There is no corporal punishment. The punishments are light : permission to go home on the holidays withdrawn ; locked up, with lesson to learn. If a boy becomes unmanageable, expulsion from the Institution is resorted to. With regard to those maintained at the expense of the Government, their punishments consist in being excluded from the free education, and charged full expenses. If that does not work they are also expelled. On the whole, however, Russian boys are very tractable. Such pupils who like to indulge in smoking are freely permitted to do so.

When a pupil has finished his course at the Institution, *i.e.,* passed his four and half years' residence to the satisfaction of the Professors, he is sent out as an assistant to some good bailiff or landowner, provided with clothing, and 125 roobles (about £15 12s. 6d.) for his expenses. If the bailiff be satisfied with him after a twelvemonth's trial, the Institution give the pupil a certificate as an Instructed Bailiff's Assistant. After six years' absence, and after having sent in a satisfactory report, on all the work he has had in hand in the meanwhile, he receives an Instructed Bailiff's Certificate; and after ten years' service, he receives a Tchin or diploma. He is bound at all times to acquaint the authorities of the Institution as to where he is engaged. A Bailiff's Assistant's salary varies between 300 and 500 roobles (about £37 10s. and £62 10s.) per annum, including board and lodging for family; a fully-empowered Bailiff or Agent draws annually 600 to 2000 roobles (about £75 to £250), together with board and lodging for family.

The above short description is, I think, sufficient to illustrate the principle of agricultural education in Russia.

The system of farming as adopted on the farm attached to the Institution is in this case an exceptional one, as will be seen at once by the rotation of crops followed :

1. Fallow, manured in June or July.
2. Rye, or winter wheat.
3. Potatoes.
4. Oats, or spring wheat ; buckwheat, or peas.
5. Clover and Timothy grass, top dressed with gypsum.
6. Clover and Timothy grass.
7. Oats.

The very fact that potatoes and clovers exist in the

rotation show that it is an exceptional one, and one far ahead of the agriculture of this part of the country. It is rather to be regretted that this is so, as in all probability very few of the scholars will for many years to come be engaged on estates labouring under the like local influence. The manager was obliged to introduce some such system in order to meet the demand of milk for the pupils, and for the soldiers quartered on Lake Kaban, and also by having easy access to the town of Kazan, where stable-manure can be bought cheap. Of course such a system is more profitable than that adopted on estates where grain is the sole product. The forty cows kept here are calved down, so as to give a sufficient quantity of milk all the year round. Allgau and Oberingthal bulls for covering are imported. The former seem to be the more favoured, as when crossed with Russian cows, a very hardy and good milker is the result. The use of such a bull is allowed to all agriculturists at a mere trifle. Considerably more calves being bred than are absolutely required, they are sold, notwithstanding their superiority, quite as cheap as ordinary Russian peasant calves: this is, however, only the case when the buyer is known to the Institution as a man who will make a good use of his purchase. Unfortunately the farm is situated on the trunk road to Orenburg, and is thus continually exposed to the ravages of the cattle-plague. The farm appears to be a very popular breeding establishment; and in order to avoid the almost annual recurrence of losses occasioned by the pest, it is now proposed at great cost to lay a new road at a considerable distance. In laying out the farm great care was exercised to have good substantial buildings; in this they have succeeded very well, and their buildings would be considered perfect even in England. Besides the good work done in stimulating cattle-breeding, they grow large quantities of seed-wheats, ryes, etc., potatoes for setting, all of which they sell to proprietors and peasants indiscriminately for the sake of introducing better crops, and thereby improving the produce of the country. Some of their seed winter-wheats find their way to Simbirsk, others to Viatka. Their potatoes are largely in demand by the peasantry, and their oats when grown on a richer soil give remarkably fine crops. As an instance of their success in this line, I may mention that at the Philadelphia Exhibition the farm received a prize medal for its linseed.

We must bear in mind that the soil is a hungry and barren sand which grows nothing without manure. Hops have been introduced; about an acre is under cultivation—it fetches a good price; cuttings are given gratis to such as wish to cultivate them. Another point, in which great interest is shown, consists in buying new implements for the sake of essaying them, so that landed proprietors may have the benefit—a benefit of which they freely avail themselves—of testing implements before laying out their money.

No doubt a model farm could never be carried on in this fashion did not the officials enter thoroughly into the spirit of the undertaking ; and it is only due to them to say that they one and all do so most heartily,—thus employing their united energies for the improvement of the agriculture of the country.

Through the kindness of Professor Malisheff I am in a position to place before the reader the actual sums required for the maintenance of the institutions in all Russia. As previously mentioned, there are five Agricultural Training Institutions throughout Russia, incurring an annual expenditure of 131,675 roobles (about £16,459), which, together with 61,505 roobles (about £7,688) for the maintenance of the farms, makes 193,180 roobles—or £24,147—spent annually on agricultural education alone.

The Government is also showing its energy in other directions. The education of the rustics being as necessary for the improvement of agriculture as that of the bailiffs, large sums are therefore annually spent on the elementary village schools, which of late years have been established in almost every commune, even in the most distant and unapproachable localities.

In the foregoing pages I have endeavoured to give a plain, unvarnished account of the Agriculture and Peasantry of Eastern Russia. To the best of my ability I have described all the more interesting details of climate, soil, cattle, etc., and have laid special stress on the measures adopted by the Government for the improvement of the country. The latter have, in consequence of the late unhappy war, been much neglected, and several matured plans which should have been put into execution last year have been laid aside until the existing clouds on the political horizon shall have cleared away.

## TABLE OF WEIGHTS AND MEASURES.

——:——

One pood (40 lbs. Russian) = 36 lbs. English.

An imperial, fiscal, or a treezataya dessyatin = 2¾ acres (approx.)

Sorokaya       „ = 3⅔    „     „

Shestdaysyattaya „ = 4      „     „

Sotelnaya      „ = 4½    „     „

A Verst    ...    ...    ...    ...    ... = ⅔ mile English.

The value of a rooble I have calculated throughout at 2s. 6d. The rooble is divided into 100 kopecks.

# APPENDIX.[1]

———

Cost of production in 1876, at Vosskresensky Zavod, near Melense, Orenburg province, on the estates of the Russia Copper Company.

*Cost per Fiscal Dessyatin.*

|  | R. | K. |
|---|---|---|
| Ploughing ... ... ... ... ... ... | 3 | — |
| Breaking down ploughed land and seeding it ... ... | 2 | 75 |
| Weeding (about 50 to 60 roobles a-year all round) ... | 0 | 05 |
| Reaping ... ... ... ... ... ... ... | 2 | 50 |
| Stacking the corn (inclusive of carrying it together) | 1 | 25 |
| Threshing (with steam threshers) and carting grain, about ten versts from field to granary, 5 kopecks per pood @ 65 poods per dessyatin ... ... ... | 3 | 25 |
| Drying the corn (part only) ... ... ... ... | — | 45 |
| Insurance of corn and farm-buildings ... ... ... | — | 60 |
| Management, general charges, and depreciation of materials ... ... ... ... ... ... | 4 | 40 |
| Total ... R. | 18 | 25 |

### NOTES.

THE soil is what may be called rather a sandy black earth.

The method of cultivation is that known as the Three-field System.

———

[1] These data and notes were sent to me by Mr. Rickard, the Superintendant in Russia, too late unfortunately to be embodied in the text.

The labour employed is from the Russian, Tartar, and Bashkeer villages of the surrounding country, contracts for for the supply of which are made six months and in some cases nearly a year beforehand. Advances are made on these contracts in the way usual in Russia. The people contracted with bring their own implements, except for the threshing, winnowing, and drying, the machinery for which the farm is provided with. It is, thanks to careful attention to this method of making contracts, that we are are able to get sufficient labour at extremely low rates. After seven years' farming the books are clear of debts on contracts, or only show arrears to a very trifling extent.

An attempt to grow winter wheat has, as everywhere else in this part of Russia, turned out to be a failure.[1]

Rye is the most reliable crop, but, in the long run, wheat is more profitable.

In the above account nothing is included for land.

---

[1] The growth of winter-wheat, wherever attempted, has hitherto been more or less a failure, and has consequently been given up. At Timashevo a two-years' trial seemed to indicate that winter-wheat might be acclimatised. Those who attempted its growth only did so for a single season ; if the experiments were continued over a series of years, using the last season's grain as seed for the next, they would most likely succeed. Unfortunately Russian landowners, in whatever they undertake, think that because a thing fails the first time, a second trial is unnecessary.—H. L. R.

www.ingramcontent.com/pod-product-compliance
Lightning Source LLC
Chambersburg PA
CBHW021822190326
41518CB00007B/698